Pandora's Hope

PANDORA'S HOPE

ESSAYS ON THE REALITY OF SCIENCE STUDIES

Bruno Latour

HARVARD UNIVERSITY PRESS
Cambridge, Massachusetts
London, England
1999

LIBRARY OF CONGRESS CATALOGING-IN-PUBLICATION DATA

Latour, Bruno.
Pandora's hope : essays on the reality of science studies / Bruno Latour
p.   cm.
Includes bibliographical references and index.
ISBN 0-674-65335-1 (alk. paper). — ISBN 0-674-65336-X (pbk. : alk. paper)
1. Realism.   2. Science—Philosophy.   I. Title
Q175.32.R42L38   1999
501—dc21   98-50061

To Shirley Strum, Donna Haraway, Steve Glickman,
and their baboons, cyborgs, and hyenas

Acknowledgments

Several chapters of this book are based on papers first published else-where. I have made no attempt to preserve the form of the originals and have gutted them whenever it was necessary for the main argument. For the sake of readers without a prior knowledge of science studies, I have kept the references to a minimum; more annotation can be found in the original publications.

I thank the editors and publishers of the following journals and books, first for having accepted my strange papers, and then for allowing them to be conjoined here: "Do Scientific Objects Have a History? Pasteur and Whitehead in a Bath of Lactic Acid," *Common Knowledge* 5, no. 1 (1993): 76–91 (translated by Lydia Davis). "Pasteur on Lactic Acid Yeast—A Partial Semiotic Analysis," *Configurations* 1, no. 1 (1993): 127–142. "On Technical Mediation," *Common Knowledge* 3, no. 2 (1994): 29–64. "Joliot: History and Physics Mixed Together," in Michel Serres, ed., *History of Scientific Thought* (London: Blackwell, 1995), 611–635. "The 'Pedofil' of Boa Vista: A Photo-Philosophical Montage," *Common Knowledge* 4, no. 1 (1995): 145–187. "Socrates' and Callicles' Settlement, or the Invention of the Impossible Body Politic," *Configurations*, 5, no. 2 (Spring 1997):189–240. "A Few Steps toward the Anthropology of the Iconoclastic Gesture," *Science in Context* 10, no. 1 (1998): 62–83.

So many people have read preliminary drafts of parts of this work that I have lost track of what is theirs and what is mine. As usual, Michel Callon and Isabelle Stengers have provided essential guidance. Behind the mask of an anonymous referee, Mario Biagioli was instrumental in giving the book its final shape. For more than ten years I have benefited from Lindsay Waters's generosity as an editor. Once again he offered a shelter for my work. My main gratitude goes, however, to John Tresch, who streamlined the language and logic of the manuscript. If readers are not satisfied with the result, they should try to imagine the jungle through which John managed to blaze this tangled trail!

I should warn the reader that this is not a book about new facts, nor it is exactly a book of philosophy. In it, using only very rudimentary tools, I simply try to present, in the space left empty by the dichotomy between subject and object, a conceptual scenography for the pair human and nonhuman. I agree that powerful arguments and detailed empirical case studies would be better, but, as sometimes happens in detective stories, a somewhat weaker, more solitary, and more adventurous strategy may succeed against the kidnapping of scientific disciplines by science warriors where others have failed.

One last caveat. Throughout this book I use the expression "science studies" as if this discipline exists and is a homogeneous body of work with a single coherent metaphysics. It would be an understatement to say that this is far from the case. Most of my colleagues disagree with my portrayal. As I do not enjoy being isolated and instead thrive on the conversations involved in a collective undertaking, I present science studies as if it is a unified field to which I belong.

Contents

CONTENTS

AUTHOR'S NOTE: Words and phrases that I use in a technical sense are marked with an asterisk; for their definitions see the Glossary.

Lucifer is the chap who brings false light . . .
I am shrouding them in the darkness of truth.

—Lakatos to Feyerabend

"Do You Believe in Reality?"

News from the Trenches of the Science Wars

"I have a question for you," he said, taking out of his pocket a crumpled piece of paper on which he had scribbled a few key words. He took a breath: "Do you believe in reality?"

"But of course!" I laughed. "What a question! Is reality something we have to believe in?"

He had asked me to meet him for a private discussion in a place I found as bizarre as the question: by the lake near the chalet, in this strange imitation of a Swiss resort located in the tropical mountains of Teresopolis in Brazil. Has reality truly become something people have to believe in, I wondered, the answer to a serious question asked in a hushed and embarrassed tone? Is reality something like God, the topic of a confession reached after a long and intimate discussion? Are there people on earth who *don't* believe in reality?

When I noticed that he was relieved by my quick and laughing answer, I was even more baffled, since his relief proved clearly enough that he had anticipated a *negative* reply, something like "Of course not! Do you think I am that naive?" This was not a joke, then: he really was concerned, and his query had been in earnest.

"I have two more questions," he added, sounding more relaxed. "Do we know more than we used to?"

"But of course! A thousand times more!"

"But is science cumulative?" he continued with some anxiety, as if he did not want to be won over too fast.

"I guess so," I replied, "although I am less positive on this one, since the sciences also forget so much, so much of their past and so much of their bygone research programs—but, on the whole, let's say yes. Why are you asking me these questions? Who do you think I am?"

I had to switch interpretations fast enough to comprehend both the monster he was seeing me as when he raised these questions and his touching openness of mind in daring to address such a monster privately. It must have taken courage for him to meet with one of these creatures that threatened, in his view, the whole establishment of science, one of these people from a mysterious field called "science studies," of which he had never before met a flesh-and-blood representative but which—at least so he had been told—was another threat to science in a country, America, where scientific inquiry had never had a completely secure foothold.

He was a highly respected psychologist, and we had both been invited by the Wenner-Grenn Foundation to a gathering made up of two-thirds scientists and one-third "science students." This division itself, announced by the organizers, baffled me. How could we be pitted *against* the scientists? That we are studying a subject matter does not mean that we are attacking it. Are biologists anti-life, astronomers anti-stars, immunologists anti-antibodies? Besides, I had taught for twenty years in scientific schools, I wrote regularly in scientific journals, I and my colleagues lived on contract research carried out on behalf of many groups of scientists in industry and in the academy. Was I not part of the French scientific establishment? I was a bit vexed to be excluded so casually. Of course I am just a philosopher, but what would my friends in science studies say? Most of them have been trained in the sciences, and several of them, at least, pride themselves on *extending* the scientific outlook to science itself. They could be labeled as members of another discipline or another subfield, but certainly not as "anti-scientists" meeting halfway with scientists, as if the two groups were opposing armies conferring under a flag of truce before returning to the battlefield!

I could not get over the strangeness of the question posed by this man I considered a colleague, yes, a colleague (and who has since become a good friend). If science studies has achieved anything, I thought, surely it has *added* reality to science, not withdrawn any from it. Instead of the stuffed scientists hanging on the walls of the armchair philosophers of science of the past, we have portrayed lively characters, immersed in their laboratories, full of passion, loaded with instruments, steeped in know-how, closely connected to a larger and more vibrant milieu. Instead of the pale and bloodless objectivity of science, we have all shown, it seemed to me, that the many nonhumans mixed into our collective life through laboratory practice have a history, flexibility, culture, blood—in short, all the characteristics that were denied to them by the humanists on the other side of the campus. Indeed, I naively thought, if scientists have a faithful ally, it is we, the "science students" who have managed over the years to interest scores of literary folk in science and technology, readers who were convinced, until science studies came along, that "science does not think" as Heidegger, one of their masters, had said.

The psychologist's suspicion struck me as deeply unfair, since he did not seem to understand that in this guerrilla warfare being conducted in the no-man's-land between the "two cultures," *we were the ones* being attacked by militants, activists, sociologists, philosophers, and technophobes of all hues, precisely because of our interest in the inner workings of scientific facts. Who loves the sciences, I asked myself, more than this tiny scientific tribe that has learned to open up facts, machines, and theories with all their roots, blood vessels, networks, rhizomes, and tendrils? Who believes more in the objectivity of science than those who claim that it can be turned into an object of inquiry?

Then I realized that I was wrong. What I would call "adding realism to science" was actually seen, by the scientists at this gathering, as a threat to the calling of science, as a way of decreasing its stake in truth and their claims to certainty. How has this misunderstanding come about? How could I have lived long enough to be asked in all seriousness this incredible question: "Do you believe in reality?" The distance between what I thought we had achieved in science studies and what was implied by this question was so vast that I needed to retrace my steps a bit. And so this book was born.

The Strange Invention of an "Outside" World

There is no natural situation on earth in which someone could be asked this strangest of all questions: "Do you believe in reality?" To ask such a question one has to become so *distant* from reality that the fear of *losing* it entirely becomes plausible—and this fear itself has an intellectual history that should at least be sketched. Without this detour we would never be able to fathom the extent of the misunderstanding between my colleague and me, or to measure the extraordinary form of radical realism that science studies has been uncovering.

I remembered that my colleague's question was not so new. My compatriot Descartes had raised it against himself when asking how an isolated mind could be *absolutely* as opposed to relatively sure of anything about the outside world. Of course, he framed his question in a way that made it impossible to give the only reasonable answer, which we in science studies have slowly rediscovered three centuries later: that we are *relatively* sure of many of the things with which we are daily engaged through the practice of our laboratories. By Descartes's time this sturdy relativism*, based on the number of *relations* established with the world, was already in the past, a once-passable path now lost in a thicket of brambles. Descartes was asking for absolute certainty from a brain-in-a-vat, a certainty that was not needed when the brain (or the mind) was firmly attached to its body and the body thoroughly involved n its normal ecology. As in Curt Siodmak's novel *Donovan's Brain,* absolute certainty is the sort of neurotic fantasy that only a surgically removed mind would look for after it had lost everything else. Like a heart taken out of a young woman who has just died in an accident and soon to be transplanted into someone else's thorax thousands of miles away, Descartes's mind requires artificial life-support to keep it viable. Only a mind put in the strangest position, looking at a world *from the inside out* and linked to the outside by nothing but the tenuous connection of the *gaze,* will throb in the constant fear of losing reality; only such a bodiless observer will desperately look for some absolute life-supporting survival kit.

For Descartes the only route by which his mind-in-a-vat could reestablish some reasonably sure connection with the outside world was through God. My friend the psychologist was thus right to phrase his query using the same formula I had learned in Sunday school: "Do you believe in reality?"—"Credo in unum Deum," or rather, "Credo in unam realitam," as my friend Donna Haraway kept chanting in Teresopolis! After Descartes, however, many people thought that going through God to reach the world was a bit expensive and far-fetched. They looked for a shortcut. They wondered whether the world could *directly* send us enough information to produce a stable image of itself in our minds.

But in asking this question the empiricists kept going along the same path. They did not retrace their steps. They never plugged the wriggling and squiggling brain back into its withering body. They were still dealing with a mind looking through the gaze at a lost outside world. They simply tried to train it to recognize patterns. God was out, to be sure, but the *tabula rasa* of the empiricists was as disconnected as the mind in Descartes's times. The brain-in-a-vat simply exchanged one survival kit for another. Bombarded by a world reduced to meaningless stimuli, it was supposed to extract from these stimuli everything it needed to recompose the world's shapes and stories. The result was like a badly connected TV set, and no amount of tuning made this precursor of neural nets produce more than a fuzzy set of blurry lines, with white points falling like snow. No shape was recognizable. Absolute certainty was lost, so precarious were the connections of the senses to a world that was pushed ever further outside. There was too much static to get any clear picture.

The solution came, but in the form of a catastrophe from which we are only now beginning to extricate ourselves. Instead of retracing their steps and taking the other path at the forgotten fork in the road, philosophers abandoned even the claim to absolute certainty, and settled instead on a makeshift solution that preserved at least some access to an outside reality. Since the empiricists' associative neural net was unable to offer clear pictures of the lost world, this must prove, they said, that the mind (still in a vat) extracts *from itself* everything it needs to form shapes and stories. Everything, that is, except the reality itself. Instead of the fuzzy lines on the poorly tuned TV set, we got the fixed tuning grid, molding the confused static, dots, and lines of the empiricist channel into a steady picture held in place by the mindset's predesigned categories. Kant's *a priori* started this extravagant form of constructivism, which neither Descartes, with his detour through God, nor Hume, with his shortcut to associated stimuli, would ever have dreamed of.

Now, with the Konigsberg broadcast, everything was ruled by the mind itself and reality came in simply to say that it was there, indeed, and not imaginary! For the banquet of reality, the mind provided the food, and the inaccessible things-in-themselves to which the world had been reduced simply dropped by to say "We are here, what you eat is not dust," but otherwise remained mute and stoic guests. If we abandon absolute certainty, Kant said, we can at least retrieve universality as long as we remain inside the restricted sphere of science, to which the world outside contributes decisively but minimally. The rest of the quest for the absolute is to be found in morality, another *a priori* certainty that the mind-in-the-vat extracts from its own wiring. Under the name of a "Copernican Revolution"* Kant invented this science-fiction nightmare: the outside world now turns around the mind-in-the-vat, which dictates most of that world's laws, laws it has extracted from itself without help from anyone else. A crippled despot now ruled the world of reality. This philosophy was thought, strangely enough, to be the deepest of all, because it had at once managed to abandon the quest for absolute certainty and to retain it under the banner of "universal *a prioris*," a clever sleight of hand that hid the lost path even deeper in the thickets.

Do we really have to swallow these unsavory pellets of textbook philosophy to understand the psychologist's question? I am afraid so, because otherwise the innovations of science studies will remain invisible. The worst is yet to come. Kant had invented a form of constructivism in which the mind-in-the-vat built everything by itself but not entirely without constraints: what it learned from itself had to be universal and could be elicited only by some experiential contact with a reality out there, a reality reduced to its barest minimum, but there nonetheless. For Kant there was still something that revolved around the crippled despot, a green planet around this pathetic sun. It would not be long before people realized that this "transcendental Ego," as Kant named it, was a fiction, a line in the sand, a negotiating position in a complicated settlement to avoid the complete loss of the world or the complete abandonment of the quest for absolute certainty. It was soon replaced by a more reasonable candidate, *society*. Instead of a mythical Mind giving shape to reality, carving it, cutting it, ordering it, it was now the prejudices, categories, and paradigms of a group of people living together that determined the representations of every one of those people. This new definition, however, in spite of the use of the word "social," had only a superficial resemblance to the realism to which we science students have become attached, and which I will outline over the course of this book.

First, this replacement of the despotic Ego with the sacred "society" did not retrace the philosophers' steps but went even *further* in distancing the individual's vision, now a "view of the world," from the definitely lost outside world. Between the two, society interposed its filters; its paraphernalia of biases, theories, cultures, traditions, and standpoints became an opaque window. Nothing of the world could pass through so many intermediaries and reach the individual mind. People were now locked not only into the prison of their own categories but into that of their social groups as well. Second, this "society" itself was just a series of minds-in-a-vat, many minds and many vats to be sure, but each of them still composed of that strangest of beasts: a detached mind gazing at an outside world. Some improvement! If prisoners were no longer in isolated cells, they were now confined to the same dormitory, the same collective mentality. Third, the next shift, from one Ego to multiple cultures, jeopardized the only good thing about Kant, that is, the universality of the *a priori* categories, the only bit of ersatz absolute certainty he had been able to retain. Everyone was not locked in the same prison any more; now there were *many* prisons, incommensurable, unconnected. Not only was the mind disconnected from the world, but each collective mind, each culture was disconnected from the others. More and more progress in a philosophy dreamed up, it seems, by prison wardens.

But there was a fourth reason, even more dramatic, even sadder, that made this shift to "society" a catastrophe following fast on the heels of the Kantian revolution. The claims to knowledge of all these poor minds, prisoners in their long rows of vats, were now made part of an even more bizarre history, were now associated with an even more ancient threat, *the fear of mob rule*. If my friend's voice quivered as he asked me "Do you believe in reality?" it was not only because he feared that all connection with the outside world might be lost, but above all because he worried that I might answer, "Reality depends on whatever the mob thinks is right at any given time." It is the resonance of these two fears, the *loss* of any certain access to reality and the *invasion* by the mob, that makes his question at once so unfair and so serious.

But before we disentangle this second threat, let me finish with the first one. The sad story, unfortunately, does not end here. However incredible it seems, it is possible to go even further along the wrong path, always thinking that a more radical solution will solve the problems accumulated from the past decision. One solution, or more exactly another clever sleight of hand, is to become so very pleased with the loss of absolute certainty and universal *a prioris* that one rejoices in abandoning them. Every defect of the former position is now taken to be its best quality. Yes, we have lost the world. Yes, we are forever prisoners of language. No, we will never regain certainty. No, we will never get beyond our biases. Yes, we will forever be stuck within our own selfish standpoint. Bravo! Encore! The prisoners are now gagging even those who ask them to look out their cell windows; they will "deconstruct," as they say—which means destroy in slow motion—anyone who reminds them that there was a time when they were free and when their language bore a connection with the world.

Who can avoid hearing the cry of despair that echoes deep down, carefully repressed, meticulously denied, in these paradoxical claims for a joyous, jubilant, free construction of narratives and stories by people forever in chains? But even if there *were* people who could say such things with a blissful and light heart (their existence is as uncertain to me as that of the Loch Ness monster, or, for that matter, as uncertain as that of the real world would be to these mythical creatures), how could we avoid noticing that we have not moved an inch since Descartes? That the mind is still in its vat, excised from the rest, disconnected, and contemplating (now with a blind gaze) the world (now lost in darkness) from the very same bubbling glassware? Such people may be able to smile smugly instead of trembling with fear, but they are still descending further and further along the spiraling curves of the same hell. At the end of this chapter we will meet these gloating prisoners again.

In our century, though, a second solution has been proposed, one that has occupied many bright minds. This solution consists of taking only a *part* of the mind out of the vat and then doing the obvious thing, that is, offering it a body again and putting the reassembled aggregate back into relation with a world that is no longer a spectacle at which we gaze but a lived, self-evident, and unreflexive extension of ourselves. In appearance, the progress is immense, and the descent into damnation suspended, since we no longer have a mind dealing with an outside world, but a lived world to which a semi-conscious and intentional body is now attached.

Unfortunately, however, in order to succeed, this emergency operation must chop the mind into even smaller pieces. The real world, the one known by science, is left entirely to itself. Phenomenology deals only with the world-for-a-human-consciousness. It will teach us a lot about how we never distance ourselves from what we see, how we never gaze at a distant spectacle, how we are always immersed in the world's rich and lived texture, but, alas, this knowledge will be of no use in accounting for how things really are, since we will never be able to escape from the narrow focus of human intentionality. Instead of exploring the ways we can shift from standpoint to standpoint, we will always be fixed in the human one. We will hear much talk about the real, fleshy, pre-reflexive lived world, but this will not be enough to cover the noise of the second ring of prison doors slamming even more tightly shut behind us. For all its claims to overcoming the distance between subject and object—as if this distinction were something that could be overcome! as if it had not been devised so as *not* to be overcome!—phenomenology leaves us with the most dramatic split in this whole sad story: a world of science left entirely to itself, entirely cold, absolutely inhuman; and a rich lived world of intentional stances entirely limited to humans, absolutely divorced from what things are in and for themselves. A slight pause on the way down before sliding even further in the same direction.

Why not choose the opposite solution and forget the mind-in-a-vat altogether? Why not let the "outside world" invade the scene, break the glassware, spill the bubbling liquid, and turn the mind into a brain, into a neuronal machine sitting inside a Darwinian animal struggling for its life? Would that not solve all the problems and reverse the fatal downward spiral? Instead of the complex "life-world" of the phenomenologists, why not study the adaptation of humans, as naturalists have studied all other aspects of "life"? If science can invade everything, it surely can put an end to Descartes's long-lasting fallacy and make the mind a wriggling and squiggling part of nature. This would certainly please my friend the psychologist—or would it? No, because the ingredients that make up this "nature," this hegemonic and all-encompassing nature*, which would now include the human species, are the *very same ones* that have constituted the spectacle of a world viewed from inside by a brain-in-a-vat. Inhuman, reductionist, causal, law-like, certain, objective, cold, unanimous, absolute— all these expressions do not pertain to nature *as such,* but to nature viewed through the deforming prism of the glass vessel!

If there is something unattainable, it is the dream of treating nature as a homogeneous unity in order to unify the different views the sciences have of it! This would require us to ignore too many controversies, too much history, too much unfinished business, too many loose ends. If phenomenology abandoned science to its destiny by limiting it to human intention, the opposite move, studying humans as "natural phenomena," would be even worse: it would abandon the rich and controversial human history of science—and for what? The averaged-out orthodoxy of a few neurophilosophers? A blind Darwinian process that would limit the mind's activity to a struggle for survival to "fit" with a reality whose true nature would escape us forever? No, no, we can surely do better, we can surely stop the downward slide and retrace our steps, retaining both the history of humans' involvement in the making of scientific facts and the sciences' involvement in the making of human history.

Unfortunately, we can't do this, not yet. We are prevented from returning to the lost crossroads and taking the other path by the dangerous bogeyman I mentioned earlier. It is the threat of mob rule that stops us, the same threat that made my friend's voice quake and quiver.

The Fear of Mob Rule

As I said, two fears lay behind my friend's strange question. The first one, the fear of a mind-in-a-vat losing its connection to a world outside, has a shorter history than the second, which stems from this truism: if reason does not rule, then mere force will take over. So great is this threat that any and every political expedient is used with impunity against those who are deemed to advocate force against reason. But where does this striking opposition between the camp of reason and the camp of force come from? It comes from an old and venerable debate, one that probably occurs in many places but that is staged most clearly and influentially in Plato's *Gorgias.* In this dialog, which I will examine in more detail in Chapters 7 and 8, Socrates, the true scientist, confronts Callicles, another of those monsters who must be interviewed in order to expose their nonsense, this time not on the shores of a Brazilian lake but in the agora in Athens. He tells Callicles: "You've failed to notice *how much power geometrical equality has among gods and men,* and this neglect of geometry has led you to believe that one should try to gain a *disproportionate* share of things" (508a).[1]

Callicles is an expert at disproportion, no doubt about that. "I think," he boasts in a preview of Social Darwinism, "we only have to look at nature to find evidence that it is right for better to have a greater share than worse. . . The superior person shall dominate the inferior person and have more than him" (483c–d). Might makes Right, Callicles frankly admits. But, as we shall see at the end of this book, there is a little snag. As both of the two protagonists are quick to point out, there are at least two sorts of Mights to consider: that of Callicles and that of the Athenian mob. "What else do you think I've been saying?" Callicles asks. "Law consists of the statements made by an assembly of slaves and assorted other forms of human debris who could be completely *discounted if it weren't for the fact they do have physical strength at their disposal*" (489c). So the question is not simply the opposition of force and reason, Might and Right, but the Might of the solitary patrician against the superior force of the crowd. How can the combined forces of the people of Athens be nullified? "Here's your position, then," Socrates ironizes: "a single clever person is almost bound to be *superior to ten thousand fools;* political power should be his and they should be his subjects; and it is appropriate for someone with political power to have more than his subjects" (490a). When Callicles speaks of brute force, what he means is an inherited moral force superior to that of ten thousand brutes.

But is it fair for Socrates to practice irony on Callicles? What sort of disproportion is Socrates himself setting in motion? What sort of power is he trying to wield? The Might that Socrates sides with is the *power of reason,* "the power of geometrical equality," the force which "rules over gods and men," which he knows, which Callicles and the mob ignore. As we shall see, there is a second little snag here, because

1. I use the recent translation by Robin Waterfield (Oxford: Oxford University Press, 1994).

there are two forces of reason, one directed against Callicles, the ideal foil, and the other directed sideways, aimed at reversing the balance of power between Socrates and all the other Athenians. Socrates is also looking for a force able to nullify that of "ten thousand fools." He too tries to get the biggest share. His success at reversing the balance of forces is so extraordinary that he boasts, at the end of the *Gorgias*, of being "the only real statesman of Athens," the only winner of the biggest share of all, an eternity of glory that will be awarded to him by Rhadamantes, Aeacus, and Minos, who preside over the tribunal of hell! He ridicules all the famous Athenian politicians, Pericles included, and he alone, equipped with "the power of geometrical equality," will rule over the citizens of the city even beyond death. One of the first of many in the long literary history of mad scientists.

"As if your slapdash history of modern philosophy is not enough," the reader may complain, "do you also have to drag us all the way back to the Greeks just to account for the question asked by your psychologist in Brazil?" I am afraid both of these detours were necessary, because only now can the two threads, the two threats, be tied together to explain my friend's worries. Only after these digressions can my position, I hope, be clarified at last.

Why, in the first place, did we even need the idea of an *outside world* looked at through a gaze from the very uncomfortable observation post of a mind-in-a-vat? This has puzzled me ever since I started in the field of science studies almost twenty-five years ago. How could it be so important to maintain this awkward position, in spite of all the cramps it gave philosophers, instead of doing the obvious: retracing our steps, pruning back the brambles hiding the lost fork in the road, and firmly walking on the other, forgotten path? And why burden this solitary mind with the impossible task of finding absolute certainty instead of plugging it into the connections that would provide it with all the relative certainties it needed to know and to act? Why shout out of both sides of our mouths these two contradictory orders: "Be absolutely disconnected!" "Find absolute proof that you are connected!" Who could untangle such an impossible double bind? No wonder so many philosophers wound up in asylums. In order to justify such a self-inflicted, maniacal torture, we would have to be pursuing a loftier goal, and such indeed has been the case. This is the place where the two threads connect: it is in order to avoid the inhuman crowd that we need to rely on another inhuman resource, the objective object untouched by human hands.

To avoid the threat of a mob rule that would make everything lowly, monstrous, and inhuman, we have to depend on something that has no human origin, no trace of humanity, something that is purely, blindly, and coldly outside of the City. The idea of a completely *outside* world dreamed up by epistemologists is the only way, in the eyes of moralists, to avoid falling prey to mob rule. *Only inhumanity will quash inhumanity.* But how is it possible to imagine an outside world? Has anyone ever seen such a bizarre oddity? No problem. We will make the world into a spectacle seen *from* the inside.

To obtain such a contrast, we will imagine that there is a mind-in-a-vat that is totally disconnected from the world and accesses it only through one narrow, artificial conduit. This minimal link, psychologists are confident, will be enough to keep the world outside, to keep the mind informed, provided we later manage to rig up some absolute means of getting certainty back—no mean feat, as it turns out. But this way we will achieve our overarching agenda: *to keep the crowds at bay.* It is because we want to fend off the irascible mob that we need a world that is totally outside—while remaining accessible!—and it is in order to reach this impossible goal that we came up with the extraordinary invention of a mind-in-a-vat disconnected from everything else, striving for absolute truth, and, alas, failing to get it. As we can see in Figure 1.1, *epistemology, morality, politics, and psychology go hand in hand and are aiming at the same settlement*.*

This is the argument of this book. It is also the reason the reality of science studies is so difficult to locate. Behind the cold epistemological question—can our representations capture with some certainty stable features of the world out there?—the second, more burning anxiety is always lurking: can we find a way to fend off the people? Conversely, behind any definition of the "social" is the same worry: will we still be able to use objective reality to shut the mob's too many mouths?

My friend's question, on the shore of the lake, shaded by the chalet's roof from the tropical noontime sun in this austral winter, becomes clear at last: "Do you believe in reality?" means "Are you willing to accept this settlement of epistemology, morality, politics, and psychology?"—to which the quick and laughing answer is, obviously: "*No!* Of course not! Who do you think I am? How could I believe real-

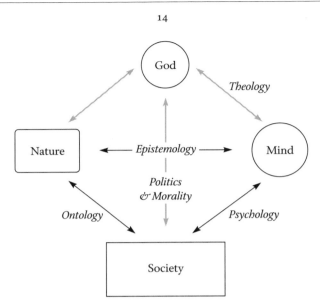

Figure 1.1 The modernist settlement. For science studies there is no sense in talking independently of epistemology, ontology, psychology, and politics—not to mention theology. In short: "out there," "nature"; "in there," the mind; "down there," the social; "up there," God. We do not claim that these spheres are cut off from one another, but rather that they all pertain to the same settlement, a settlement that can be replaced by several alternative ones.

ity to be the answer to a question of belief asked by a brain-in-a-vat terrified of losing contact with an outside world because it is even more terrified of being invaded by a social world stigmatized as inhuman?" Reality is an object of belief only for those who have started down this impossible cascade of settlements, always tumbling into a worse and more radical solution. Let them clean up their own mess and accept the responsibility for their own sins. My trajectory has always been different. "Let the dead bury the dead," and, please, listen for one minute to what we have to say on our own account, instead of trying to shut us up by putting in our mouths the words that Plato, all those centuries ago, placed in the mouths of Socrates and Callicles to keep the people silent.

Science studies, as I see it, has made two related discoveries that were very slow in coming because of the power of the settlement that I have now exposed—as well as for a few other reasons I will explain later. This joint discovery is that *neither the object nor the social* has the *inhuman* character that Socrates' and Callicles' melodramatic show required. When we say there is no outside world, this does not mean that we deny its existence, but, on the contrary, that we refuse to grant it the ahistorical, isolated, inhuman, cold, objective existence that it was given *only* to combat the crowd. When we say that science is social, the word social for us does not bear the stigma of the "human debris," of the "unruly mob" that Socrates and Callicles were so quick to invoke in order to justify the search for a force strong enough to reverse the power of "ten thousand fools."

Neither of these two monstrous forms of inhumanity—the mob "down there," the objective world "out there"—interests us very much. And thus we have no need for a mind- or brain-in-a-vat, that crippled despot constantly fearful of losing either "access" to the world or its "superior force" against the people. We long neither for the absolute certainty of a contact with the world nor for the absolute certainty of a transcendent force against the unruly mob. We do not *lack* certainty, because we never dreamed of *dominating* the people. For us there is no inhumanity to be quashed with another inhumanity. Humans and nonhumans are enough for us. We do not need a social world to break the back of objective reality, nor an objective reality to silence the mob. It is quite simple, even though it may sound incredible in these times of the science wars: we are *not* at war.

As soon as we refuse to engage the scientific disciplines in this dispute about who should hold sway over the people, the lost crossroads is rediscovered, and there is no major difficulty in treading along the neglected path. Realism now returns in force, as will be made obvious, I hope, in later chapters, which should look like milestones along the route to a more "realistic realism." My argument in this book recapitulates the halting "two steps forward, one step back" advance of science studies along this long-forgotten pathway.

We started when we first began to talk about scientific *practice* * and thus offered a more realistic account of science-in-the-making, grounding it firmly in laboratory sites, experiments, and groups of colleagues, as I do in Chapters 2 and 3. Facts, we found, were clearly fabricated. Then realism gushed forth again when, instead of talking about objects and objectivity, we began to speak of *nonhumans* * that were socialized through the laboratory and with which scientists and engineers began to swap properties. In Chapter 4 we see how Pasteur makes his microbes while the microbes "make their Pasteur"; Chapter 6 offers a more general treatment of humans and nonhumans folding into each other, forming constantly changing collectives. Whereas objects had been made cold, asocial, and distant for political reasons, we found that nonhumans were close, hot, and easier to enroll and to enlist, adding more and more reality to the many struggles in which scientists and engineers had engaged.

But realism became even more abundant when nonhumans began to have a *history*, too, and were allowed the multiplicity of interpretations, the flexibility, the complexity that had been reserved, until then, for humans (see Chapter 5). Through a series of counter-Copernican revolutions*, Kant's nightmarish fantasy slowly lost its pervasive dominance over the philosophy of science. There was again a clear sense in which we could say that words have *reference* to the world and that science grasps the things themselves (see Chapters 2 and 4). Naïveté was back at last, a naïveté appropriate for those who had never understood how the world could be "outside" in the first place. We have yet to provide a real alternative to that fateful distinction between construction and reality; I attempt to provide one here with the notion of "factish." As we see in Chapter 9, "factish" is a combination of the words "fact" and "fetish," in which the work of fabrication has been twice added, canceling the twin effects of belief and knowledge.

Instead of the three poles—a reality "out there," a mind "in there," and a mob "down there"—we have finally arrived at a sense of what I call a *collective**. As the explication of the *Gorgias* in Chapters 7 and 8 demonstrates, Socrates has defined this collective very well before switching to his bellicose collusion with Callicles: "The expert's opinion is that co-operation, love, order, discipline, and justice *bind* heaven and earth, gods and men. That's why they call the universe an *ordered whole*, my friend, rather than a disorderly mess or an *unruly shambles*" (507e–508a).

Yes, we live in a hybrid world made up at once of gods, people, stars, electrons, nuclear plants, and markets, and it is our duty to turn it into either an "unruly shambles" or an "ordered whole," a *cosmos* as the Greek text puts it, undertaking what Isabelle Stengers gives the beautiful name of cosmopolitics* (Stengers 1996). Once there is no longer a mind-in-a-vat looking through the gaze at an outside world, the search for absolute certainty becomes less urgent, and thus there is no great difficulty in reconnecting with the relativism, the relations, the relativity on which the sciences have always thrived. Once the social realm no longer bears these stigmata branded upon it by those who want to silence the mob, there is no great difficulty in recognizing the human character of scientific practice, its lively history, its many connections with the rest of the collective. Realism comes back like blood through the many vessels now reattached by the clever hands of the surgeons—there is no longer any need for a survival kit. After following this route, no one would even think of asking the bizarre question "Do you believe in reality?"—at least not of asking *us!*

The Originality of Science Studies

Nevertheless, my friend the psychologist would still be entitled to pose another, more serious query: "Why is it that, in spite of what you claim your field has achieved, I was *tempted* to ask you my silly question *as if* it were a worthwhile one? Why is it that in spite of all these philosophies you zigzagged me through, I still doubt the radical realism you advocate? I can't avoid the nasty feeling that there is a science war going on. In the end, are you a friend of science or its enemy?"

Three different phenomena explain, to me at least, why the novelty of "science studies" cannot be registered so easily. The first is that we are situated, as I said, in the no-man's-land between the two cultures, much like the fields between the Siegfried and Maginot lines in which French and German soldiers grew cabbages and turnips during the "phony war" in 1940. Scientists always stomp around meetings talking about "bridging the two-culture gap," but when scores of people from outside the sciences begin to build just that bridge, they recoil in horror and want to impose the strangest of all gags on free speech since Socrates: only scientists should speak about science!

Just imagine if that slogan were generalized: only politicians should speak about politics, businessmen about business; or even worse: only rats will speak about rats, frogs about frogs, electrons about electrons! Speech implies by definition the risk of misunderstanding across the huge gaps between different species. If scientists want to bridge the two-culture divide for good, they will have to get used to a lot of noise and, yes, more than a little bit of nonsense. After all, the humanists and the literati do not make such a fuss about the many absurdities uttered by the team of scientists building the bridge from the other end. More seriously, bridging the gap cannot mean extending the unquestionable *results* of science in order to stop the "human debris" from behaving irrationally. Such an attempt can at best be called pedagogy, at worst propaganda. This cannot pass for the cosmopolitics that would require the collective to socialize into its midst the humans, the nonhumans, and the gods together. Bridging the two-culture gap cannot mean lending a helping hand to Socrates' and Plato's dreams of utter control.

But where does the two-culture debate itself originate? In a division of labor between the two sides of the campus. One camp deems the sciences accurate only when they have been purged of any contamination by subjectivity, politics, or passion; the other camp, spread out much more widely, deems humanity, morality, subjectivity, or rights worthwhile only when they have been protected from any contact with science, technology, and objectivity. We in science studies fight against these two purges, against both purifications at once, and this is what makes us traitors to both camps. We tell the scientists that *the more connected a science* is to the rest of the collective, *the better* it is, the more accurate, the more verifiable, the more solid (see Chapter 3)—and this runs against all the conditioned reflexes of epistemologists. When we tell them that the social world is good for science's health, they hear us as saying that Callicles' mobs are coming to ransack their laboratories.

But, against the other camp, we tell the humanists that *the more nonhumans share existence with humans, the more humane* a collective is—and this too runs against what they have been trained for years to believe. When we try to focus their attention on solid facts and hard mechanisms, when we say that objects are good for the subjects' health because objects have none of the inhuman characteristics they fear so much, they scream that the iron hand of objectivity is turning frail and pliable souls into reified machines. But we keep defecting and counter-defecting from both sides, and we insist and insist again that there is a social history of things and a "thingy" history of humans, but that neither "the social" nor "the objective world" plays the role assigned to it by Socrates and Callicles in their grotesque melodrama.

If anything, and here we can be rightly accused of a slight lack of symmetry, "science students" fight the humanists who are trying to invent a human world purged of nonhumans *much more* than we combat the epistemologists who are trying to purify the sciences of any contamination by the social. Why? Because scientists spend only a fraction of their time purifying their sciences and, frankly, do not give a damn about the philosophers of science coming to their rescue, while the humanists spend all their time on and take very seriously the task of freeing the human subjects from the dangers of objectification and reification. Good scientists enlist in the science wars only in their spare time or when they are retired or have run out of grant money, but the others are up in arms day and night and even get granting agencies to join in their battle. This is what makes us so angry about the suspicion of our scientist colleagues. They don't seem to be able to differentiate friends from foes anymore. Some are pursuing the vain dream of an autonomous and isolated science, Socrates' way, while we are pointing out the very means they need to reconnect the facts to the realities without which the existence of the sciences cannot be sustained. Who first offered us this treasure trove of knowledge? The scientists themselves!

I find this blindness all the more bizarre because, in the last twenty years, many scientific disciplines have joined us, crowding into the tiny no-man's-land between the two lines. This is the second reason "science studies" is so contentious. By mistake, it is caught in the middle of another dispute, this one *within* the sciences themselves. On one side there are what could be called the "cold war disciplines," which still look superficially like the Science of the past, autonomous and detached from the collective; on the other side there are strange imbroglios of politics, science, technology, markets, values, ethics, facts, which cannot easily be captured by the word Science with a capital S.

If there is some plausibility in the assertion that cosmology does not have the slightest connection with society—although even that is wrong, as Plato reminds us so tellingly—it is hard to say the same of neuropsychology, sociobiology, primatology, computer sciences, marketing, soil science, cryptology, genome mapping, or fuzzy logic, to name just a few of these active zones, a few of the "disorderly messes" as Socrates would call them. On the one hand we have a model that still applies the earlier slogan—the less connected a science the better—while on the other we have many disciplines, uncertain of their exact status, striving to apply the old model, unable to reinstate it, and not yet prepared to mutter something like what we have been saying all along: "Relax, calm down, the more connected a science is the better. Being part of a collective will not deprive you of the nonhumans you socialize so well. It will only deprive you of the po-lemical kind of objectivity that has no other use than as a weapon for waging a political war *against* politics."

To put it even more bluntly, science studies has become a hostage in a huge shift from Science to what we could call Research (or Science No. 2, as I will call it in Chapter 8). While Science had certainty, cold-ness, aloofness, objectivity, distance, and necessity, Research appears to have all the opposite characteristics: it is uncertain; open-ended; immersed in many lowly problems of money, instruments, and know-how; unable to differentiate as yet between hot and cold, subjective and objective, human and nonhuman. If Science thrived by behaving as if it were totally disconnected from the collective, Research is best seen as a *collective experimentation* about what humans and nonhumans together are able to swallow or to withstand. It seems to me that the second model is wiser than the former. No longer do we have to choose between Right and Might, because there is now a third party in the dispute, that is, the collective*; no longer do we have to decide between Science and Anti-Science, because here too there is a third party—*the same* third party, the collective.

Research is this zone into which humans and nonhumans are thrown, in which has been practiced, over the ages, the most extraor-dinary collective experiment to distinguish, in real time, between "cosmos" and "unruly shambles" with no one, neither the scientists nor the "science students," knowing in advance what the provisional answer will be. Maybe science studies is anti-Science, after all, but in that case it is wholeheartedly *for* Research, and, in the future, when the spirit of the times will have taken a firmer grip on public opinion, it will be in the same camp as all of the active scientists, leaving on the other side only a few disgruntled cold-war physicists still wishing to help Socrates shut the mouths of the "ten thousand fools" with an un-questionable and indisputable absolute truth coming from nowhere. The opposite of relativism, we should never forget, is called absolut-ism (Bloor [1976] 1991).

I am being a bit disingenuous, I know—because there is a third reason that makes it hard to believe that science studies could have so many goodies to offer. By an unfortunate coincidence, or maybe through a strange case of Darwinian mimicry in the ecology of the social sciences, or—who knows?—through some case of mutual contamination, science studies bears a superficial resemblance to those prisoners locked in their cells whom we left, a few pages ago, in their slow descent from Kant to hell and smiling smugly all the way down, since they claim no longer to care about the ability of language to refer to reality. When we talk about hybrids and imbroglios, mediations, practice, networks, relativism, relations, provisional answers, partial connections, humans and nonhumans, "disorderly messes," it may sound as if we, too, are marching along the same path, in a hurried flight from truth and reason, fragmenting into ever smaller pieces the categories that keep the human mind forever removed from the presence of reality. And yet—there is no need to paper it over—just as there is a fight inside the scientific disciplines between the model of Science and the model of Research, there is a fight in the social sciences and the humanities between two opposite models, one that can loosely be called postmodern* and the other that I have called nonmodern*. Everything the first takes to be a justification for more absence, more debunking, more negation, more deconstruction, the second takes as a proof of presence, deployment, affirmation, and construction.

The cause of the radical differences as well as of the passing resemblances is not difficult to ferret out. Postmodernism, as the name indicates, is descended from the series of settlements that have defined modernity. It has inherited from these the disconnected mind-in-the-vat's quest for absolute truth, the debate between Might and Right, the radical distinction between science and politics, Kant's constructivism, and the critical urge that goes with it, but it has *stopped* believing it is possible to carry out this implausible program successfully. In this disappointment it shows good common sense, and that is something to say in its favor. But it has not retraced the path of modernity all the way back to the various bifurcations that started this impossible project in the first place. It feels the same nostalgia as modernism, except that it tries to take on, as positive features, the overwhelming failures of the rationalist project. Hence its apology on behalf of Callicles and the Sophists, its rejoicing in virtual reality, its debunking of "master narratives," its claim that it is good to be stuck inside one's own standpoint, its overemphasis on reflexivity, its maddening efforts to write texts that do not carry any risk of presence.

Science studies, as I see it, has been engaged in a very different nonmodern task. For us, modernity has never been the order of the day. Reality and morality have never been lacking. The fight for or against absolute truth, for or against multiple standpoints, for or against social construction, for or against presence, has never been the important one. The program of debunking, exposing, avoiding being taken in, steals energy from the task that has always seemed much more important to the collective of people, things, and gods, namely, the task of sorting out the "cosmos" from an "unruly shambles." We are aiming at a *politics of things*, not at the bygone dispute about whether or not words refer to the world. Of course they do! You might as well ask me if I believe in Mom and apple pie or, for that matter, if I believe in reality!

Are you still unconvinced, my friend? Still uncertain if we are fish or fowl, friends or foes? I must confess that it takes more than a small act of faith to accept this portrayal of our work in such a light, but since you asked your question with such an open mind, I thought you deserved to be answered with the same frankness. It is true that it is a bit difficult to locate us in the middle of the two-culture divide, in the midst of the epochal shift from Science to Research, torn between the postmodern and the nonmodern predicament. I hope you are convinced, at least, that there is no deliberate obfuscation in our position, but that being faithful to your own scientific work in these troubled times is just damned difficult. In my view, your work and that of your many colleagues, your effort to establish facts, has been taken hostage in a tired old dispute about how best to control the people. We believe the sciences deserve better than this kidnapping by Science.

Contrary to what you may have thought when you asked me for this private conversation, far from being the ones who have limited science to "mere social construction" by the frantic disorderly mob invented to satisfy Callicles' and Socrates' urge for power, we in science studies may be *the first to have found a way to free the sciences from politics*— the politics of reason, that old settlement among epistemology, morality, psychology, and theology. We may be the first to have freed nonhumans from the politics of objectivity and humans from the politics of subjectification. The disciplines themselves, the facts and the artifacts with their beautiful roots, their delicate articulations, their many tendrils, and their fragile networks remain, for the most part, to be investigated and described. I try my best, in the pages that follow, to untangle a few of them. Far from the rumblings of the science wars in which neither you nor I want to fight (well, maybe I won't mind firing a few shots!), facts and artifacts can be part of many other conversations, much less bellicose, much more productive, and, yes, much friendlier.

I have to admit I am being disingenuous again. In opening the black box of scientific facts, we knew we would be opening Pandora's box. There was no way to avoid it. It was tightly sealed as long as it remained in the two-culture no-man's-land, buried among the cabbages and the turnips, blissfully ignored by the humanists trying to avoid all the dangers of objectification and by the epistemologists trying to fend off all the ills carried by the unruly mob. Now that it has been opened, with plagues and curses, sins and ills whirling around, there is only one thing to do, and that is to go even deeper, all the way down into the almost-empty box, in order to retrieve what, according to the venerable legend, has been left at the bottom—yes, *hope*. It is much too deep for me on my own; are you willing to help me reach it? May I give you a hand?

Circulating Reference

Sampling the Soil in the Amazon Forest

The only way to understand the reality of science studies is to follow what science studies does best, that is, paying close attention to the details of scientific practice. Once we have described this practice from up close as other anthropologists do when they go off to live among foreign tribes, we will be able to raise again the classic question that the philosophy of science attempted to solve without the help of an empirical grounding: how do we pack the world into words? To begin with I have chosen a discipline, soil science, and a situation, a field trip in the Amazon, that will not require too much previous knowledge. As we examine in detail the practices that produce information about a state of affairs, it should become clear how very unrealistic most of the philosophical discussions about realism have been.

The old settlement started from a gap between words and the world, and then tried to construct a tiny footbridge over this chasm through a risky correspondence between what were understood as totally different ontological domains—language and nature. I want to show that there is neither correspondence, nor gaps, nor even two distinct ontological domains, but an entirely different phenomenon: circulating reference*. To capture it, we need to slow our pace a bit and set aside all our time-saving abstractions. With the help of my camera, I will attempt to bring some sort of order to the jungle of scientific practice. Let us turn now to the first freeze-frame of this photo-philosophical montage. If a picture is worth a thousand words, a map, as we shall see, can be worth a whole forest.

On the left in Figure 2.1 is a large savanna. On the right abruptly be-

Figure 2.1

gins the outskirts of a dense forest. One side is dry and empty, the other wet and teeming with life, and though it may look as if local inhabitants have created this edge, no one has ever cultivated these lands and no line has traced the border, which extends for hundreds of kilometers. Although the savanna serves as a pasture for some landowners' cattle, its limit is the natural edge of the forest, not a man-made boundary.

Little figures lost in the landscape, pushed off to the side as in a painting by Poussin, point at interesting phenomena with their fingers and pens. The first character, pointing at some trees and plants, is Edileusa Setta-Silva. She is Brazilian. She lives in this region, teaching botany at the small university in the little town of Boa Vista, the capital of the Amazonian province of Roraima. Just to her right another person looks on attentively, smiling at what Edileusa is showing him. Armand Chauvel is from France. He has been sent on this trip by ORSTOM, the research institute of the French former colonial empire, the "agency for the development of cooperative scientific research."

Armand is not a botanist but a pedologist (pedology is one of the soil sciences, not to be confused with either geology, the science of subsoil, or podiatry, the medical art of treating feet); he resides about a thousand kilometers away in Manaus, where ORSTOM finances his laboratory in a Brazilian research center known as INPA.

The third person, taking notes in a small notebook, is Héloïsa Filizola. She is a geographer, or rather, as she insists, a geomorphologist, studying the natural and social history of the shape of the land. She is Brazilian like Edileusa, but from the south, from São Paulo, which is thousands of kilometers away, almost another country. She is also a professor at a university, though one far larger than the one in Boa Vista.

As for me, I'm the one taking this picture and describing this scene. My job as a French anthropologist is to follow these three at work. Familiar with laboratories, I decided for a change to observe a field expedition. I also decided, being something of a philosopher, to use my report on the expedition as a chance to study empirically the epistemological question of scientific reference. Through this photo-philosophical account I will bring before your eyes, dear reader, a small part of the forest of Boa Vista; I will show you some traits of my scientists' intelligence; and I will strive to make you aware of the labor required for this transport and that reference.

What are they talking about on this early morning in October 1991, after driving the jeep over terrible roads to reach this field site, which for many years now Edileusa has been carefully dividing into sections, where she has been noting the growth patterns of the trees and the sociology and demography of the plants? They are talking about the soil and the forest. Yet because they belong to two very different disciplines, they speak of them in different ways.

Edileusa is pointing to a species of fire-resistant trees that usually grow only in the savanna and that are surrounded by many small seedlings. Yet she has also found trees of this same species along the edge of the forest, where they are more vigorous but do not shade any smaller plants. To her surprise she has even managed to find a few of these trees ten meters into the forest, where they tend to die from insufficient light. Might the forest be advancing? Edileusa hesitates. For her, the large tree that you see in the background of this picture may be a scout sent by the forest as an advance guard, or perhaps a rear guard, sacrificed by the retreating forest to the merciless encroachment of the savanna. Is the forest advancing like Birnam Wood toward Dunsinane, or is it retreating?

This is the question that interests Armand; this is why he has come from so far away. Edileusa believes the forest is advancing, but she cannot be certain because the botanical evidence is confused: the same tree may be playing either of two contradictory roles, scout or rear guard. For Armand, the pedologist, at first glance it is the savanna that must be eating up the forest little by little, degrading the clay soil necessary for healthy trees into a sandy soil in which only grass and small shrubs can survive. If all her knowledge as a botanist makes Edileusa side with the forest, all his knowledge of pedology makes Armand lean toward the savanna. Soil goes from clay to sand, not from sand to clay—everyone knows that. Soil cannot avoid degradation; if the laws of pedology do not make this clear, then the laws of thermodynamics should.

Thus our friends are faced with an interesting cognitive and disciplinary conflict. A field expedition to resolve it was easy to justify. The entire world is interested in the Amazon forest. The news that the Boa Vista forest, on the outskirts of dense tropical zones, is advancing or retreating should indeed be of interest to businessmen. It was equally easy to justify mixing the know-how of botany with that of pedology in a single expedition, even though such a combination is unusual. The chain of translation* that allows them to obtain funding is not very long. I will not deal at length with the politics surrounding this expedition, since in this chapter I wish to concentrate on scientific reference as a philosopher, not on its "context" as a sociologist. (I apologize in advance to the reader, because I am going to omit many aspects of this field trip that pertain to the colonial situation. What I want to do here is to mimic as much as possible the problems and vocabulary of the philosophers in order to rework the question of reference. Later I will rework the notion of context, and in Chapter 3 I will correct the distinction between content and context.)

In the morning before leaving we meet on the terrace of the little hotel restaurant called Eusebio (Figure 2.2). We are in the center of Boa Vista, a rather rough frontier town where the *garimperos* sell the gold that they have extracted by shovel, by mercury, by gun, from the forest and from the Yanomami.

For this expedition, Armand (on the right) has asked for the help

Figure 2.2

of his colleague René Boulet (the man with the pipe). French like Armand, René is also a pedologist from ORSTOM but based in São Paulo. Here are two men and two women. Two Frenchmen and two Brazilians. Two pedologists, one geographer, and a botanist. Three visitors and one "native." All four are leaning over two kinds of maps and pointing at the precise location of the site marked out by Edileusa. Also on the table is an orange box, the indispensable topofil, which I will discuss later.

The first map, printed on paper, corresponds to the section of the atlas, compiled by Radambrasil on a scale of one to one million, that covers all of Amazonia. I will soon learn to put quotation marks around the word "covers," since, according to my informants, the beautiful yellow, orange, and green colors on the map do not always correspond to the pedological data. This is the reason they wish to zoom in, using black-and-white aerial photographs on a scale of one to fifty thousand. A single inscription* would not inspire trust, but the superposition of the two allows at least a quick indication of the exact location of the site.

This is a situation so trivial that we tend to forget its novelty: here are four scientists whose gaze is able to dominate two maps of the very landscape that surrounds them. (Both of Armand's hands and Edileusa's right hand must continually smooth out the corners of the map, otherwise the comparison would be lost and the feature they are trying to find would not appear.) Remove both maps, confuse cartographic conventions, erase the tens of thousands of hours invested in Radambrasil's atlas, interfere with the radar of planes, and our four scientists would be lost in the landscape and obliged once more to begin all the work of exploration, reference marking, triangulation, and squaring performed by their hundreds of predecessors. Yes, scientists master the world, but only if the world comes to them in the form of two-dimensional, superposable, combinable inscriptions*. It has always been the same story, ever since Thales stood at the foot of the Pyramids.

Note, dear reader, that the owner of the restaurant seems to have the same problem as our researchers and Thales. If the owner had not written the number 29 in big black letters on the table on the terrace, he would be unable to navigate his own restaurant; without such markings he would not be able to keep track of the orders or distribute the bills. He looks like a mafioso as he lowers his enormous belly into a chair when he arrives in the morning, but he, even he, needs inscriptions to oversee the economy of his small world. Erase the numbers inscribed on the table, and he would be as lost in his restaurant as our scientists would be in the forest without maps.

In the previous picture our friends were immersed in a world in which distinct features could be discerned only if pointed out with a finger. Our friends fumbled. They hesitated. But in this picture they are sure of themselves. Why? Because they can point with their fingers to phenomena taken in by the eye and susceptible to the know-how of their age-old disciplines: trigonometry, cartography, geography. In accounting for knowledge thus acquired, we should not forget to mention the rocket ship Ariane, orbiting satellites, data banks, draftspeople, engravers, printers, and all those whose work here manifests itself as paper. There remains that gesture of the finger, the "index" par excellence. "Here, there, I, Edileusa, I leave words behind and I designate, on the map, on the restaurant table, the location of the site where we will go later, when Sandoval the technician comes to get us in the jeep."

How does one pass from the first image to the second—from igno-rance to certainty, from weakness to strength, from inferiority in the face of the world to the domination of the world by the human eye? These are the questions that interest me, and for which I have traveled so far. Not to resolve, as my friends intend, the dynamic of the forest-savanna transition, but to describe the tiny gesture of a finger pointed toward the *referent of discourse*. Do the sciences speak of the world? This is what they claim, and yet Edileusa's finger designates a single coded point on a photograph that bears a mere resemblance, in certain traits, to figures printed on the map. At the restaurant table we are quite distant from the forest, yet she talks about it with assurance, as if she had it under her hand. The sciences do not speak of the world but, rather, construct representations that seem always to push it away, but also to bring it closer. My friends want to discover whether the forest advances or recedes, and I want to know how the sciences can be at the same time realist and constructivist, immediate and intermediary, reliable and fragile, near and far. Does the discourse of science have a referent? When I speak of Boa Vista, to what does the spoken word re-fer? Do science and fiction differ? And one additional query: how does my way of talking about this photomontage differ from the man-ner in which my informants speak of their soil?

Laboratories are excellent sites in which to understand the produc-tion of certainty, and that is why I enjoy studying them so much, but like these maps, they have the major disadvantage of relying on the indefinite sedimentation of other disciplines, instruments, languages, and practices. One no longer sees science stammer, making its debut, creating itself from nothing in direct confrontation with the world. In the laboratory there is always a preconstructed universe that is mi-raculously similar to that of the sciences. In consequence, since the known world and the knowing world are always performing in con-cert with each other, reference always resembles a tautology (Hacking 1992). But not in Boa Vista, or so it seems. Here science does not blend well with the *garimperos* and the white waters of the Rio Branco. What luck! In accompanying this expedition I will be able to follow the trail of a relatively poor and weak discipline that will, before my eyes, take its first steps, just as I would have been able to observe the teeterings and totterings of geography had I, in past centuries, run through Brazil after Jussieu or Humboldt.

Here in the great forest (Figure 2.3), a horizontal branch is fore-

Figure 2.3

grounded against an otherwise uniformly green background. On this branch, attached to a rusty nail, is a little tin tag on which is written the number 234.

In the thousands of years in which humans have traveled through this forest, slashing and burning in order to cultivate it, no one had ever before had the peculiar idea of attaching numbers to it. It took a scientist, or perhaps a forester designating trees to be felled. In either case, this numbering of trees is, we must assume, the work of a meticulous bookkeeper (Miller 1994).

After an hour in the jeep, we have arrived at the plot of land that Edileusa has been charting for many years. Like the owner of the restaurant in the previous picture, she would not be able to remember the differences between patches of the forest for very long without marking them in some way. She has therefore placed tags at regular intervals so as to cover the few hectares of her field site in a grid of Cartesian coordinates. These numbers will allow her to register the variations of growth and the emergence of species in her notebook. Each plant possesses what is called a reference, both in geometry (through the attribution of coordinates) and in the management of stock (through the affixing of specific numbers).

Despite the pioneering quality of this expedition, it turns out, I am not assisting at the birth of a science *ex nihilo*. My pedological colleagues cannot fruitfully begin their work unless the site has already been marked out by *another* science, botany. I thought I was deep in the forest, but the implication of this sign, "234," is that we are *in a laboratory*, albeit a minimalist one, traced by the grid of coordinates. The forest, divided into squares, has already lent itself to the collection of information on paper that likewise takes a quadrilateral form. I rediscover the tautology that I believed I was escaping by coming into the field. One science always hides another. If I were to tear down these tree tags, or if I were to mix them up, Edileusa would panic like those giant ants whose paths I disturb by slowly passing my finger across their chemical freeways.

Edileusa cuts off her specimens (Figure 2.4). We always forget that the word "reference" comes from the Latin *referre*, "to bring back." Is the referent what I point to with my finger outside of discourse, or is it what I bring back inside discourse? The whole object of this montage is to answer that question. If I appear to be taking a roundabout route to the response, it is because there is no fast-forward button for unreeling the practice of science if I want to follow the many steps between our arrival at the site and the eventual publication.

In this frame Edileusa extracts, from the broad diversity of plants, specimens that correspond to those recognized taxonomically as *Guatteria schomburgkiana, Curatella americana,* and *Connnarus favosus.* She says she recognizes them as well as she does the members of her own family. Each plant that she removes represents thousands of the same species present in the forest, in the savanna, and on the border of the two. It is not a bouquet of flowers she is assembling but evidence that she wants to keep as a reference (using here another sense of the word). She must be able to retrieve what she writes in her notebooks

Figure 2.4

and refer to it in the future. In order to be able to say that *Afulamata diasporis*, a common forest plant, is found in the savanna but only in the shadow of a few forest plants that manage to survive there, she must preserve, not the whole population, but a sample that will serve as a silent witness for this claim.

In the bouquet she has just picked we can recognize two features of reference: on the one hand an economy, an induction, a shortcut, a funnel in which she picks one blade of grass as the sole representative of thousands of blades of grass; and on the other hand the preservation of a specimen that will later act as guarantor when she is in doubt herself or when, for various reasons, colleagues may doubt her claims.

Like the footnotes used in scholarly works to which the inquisitive or the skeptical "make reference" (yet another use of the word), this armful of specimens will guarantee the text that results from her field expedition. The forest cannot directly give its credit to Edileusa's text, but she can be credited indirectly through the extraction of a representative guarantor, neatly preserved and tagged, that can be transported, along with her notes, to her collection at the university in Boa Vista. We will be able to go from her written report to the names of the plants, from these names to the dried and classified specimens. And if there is ever a dispute, we will, with the help of her notebook, be able to go back from these specimens to the marked-out site from which she started.

A text speaks of plants. A text has plants for footnotes. A leaflet rests on a bed of leaves.

What will happen to these plants? They will be transported further, placed in a collection, a library, a museum. Let us see what will happen to them in one of these institutions, because this step is much better known and has been more often described (Law and Fyfe 1988; Lynch and Woolgar 1990; Star and Griesemer 1989; Jones and Galison 1998). Then we will focus again on the intermediary steps. In Figure 2.5 we are in a botanical institute, quite far from the forest, in Manaus. A cabinet with three ranks of shelves constitutes a work space crisscrossed in columns and rows, x- and y-axes. Each compartment shown in this photograph is used as much for classification as for tagging and preservation. This piece of furniture is a theory, only slightly heavier than the tag in Figure 2.3 but much more capable of organizing this office, a

Figure 2.5

perfect intermediary between hardware (since it shelters) and software (since it classifies), between a box and the tree of knowledge.

The tags designate the names of the collected plants. The dossiers, files, and folders shelter not text—forms or mail—but plants, the very plants that the botanist removed from the forest, that she dried in an oven at 40 degrees Celsius to kill the fungi, and that she has since pressed between newspapers.

Are we far from or near to the forest? Near, since one finds it here in the collection. The *entire* forest? No. Neither ants, nor trapdoor spiders, nor trees, nor soil, nor worms, nor the howler monkeys whose cry can be heard for miles are in attendance. Only those few specimens and representatives that are of interest to the botanist have made it into the collection. So are we, therefore, far from the forest? Let us say we are in between, possessing all of it through these delegates, as if Congress held the entire United States; a very economical metonymy in science as in politics, by which a tiny part allows the grasping of the immense whole.

And what would be the point of transporting the whole forest here? One would get lost in it. It would be hot. The botanist would in any case be unable to see beyond her small plot. Here, however, the air conditioner is humming. Here, even the walls become part of the multiple crisscrossed lines of the chart where the plants find a place that belongs to them within the taxonomy that has been standardized for many centuries. Space becomes a table chart, the table chart becomes a cabinet, the cabinet becomes a concept, and the concept becomes an institution.

Therefore we are neither very far from nor very close to the field site. We are at a good distance, and we have transported a small number of pertinent features. During the transportation something has been preserved. If I can manage to grasp this *invariant*, this *je ne sais quoi*, I believe, I will have understood scientific reference.

In this little room where the botanist shelters her collection (Figure 2.6) is a table, similar to that in the restaurant, on which the specimens brought back from distinct locations at different times are now displayed. Philosophy, the art of wonderment, should consider this table carefully, since it is where we see why the botanist gains so much more from her collection than she loses by distancing herself from the

Figure 2.6

forest. Let us first review what we know of that superiority before again attempting to follow the intermediary steps.

The first advantage: comfort. In leafing through the pages of newsprint, the researcher makes the dried stems and flowers visible so she can examine them at leisure, writing just beside them as if the stems and flowers could imprint themselves directly onto the paper or at least become compatible with the paper world. The supposedly vast distance between writing and things is now only a few centimeters.

A second advantage, just as important, is that once classified, specimens from different locations and times become contemporaries of one another on the flat table, all visible under the same unifying gaze. This plant, classified three years ago, and this other, obtained more than a thousand kilometers away, conspire on the table to form a synoptic tableau.

A third advantage, again equally decisive, is that the researcher can shift the position of specimens and substitute one for another as if shuffling cards. Plants are not exactly signs, yet they have become as mobile and recombinable as the lead monotype characters of a printing press.

Hardly surprising, then, that in the calm and cool office the botanist who patiently arranges the leaves is able to discern emerging patterns that no predecessor could see. The contrary would be much more surprising. Innovations in knowledge naturally emerge from the collection deployed on the table (Eisenstein 1979). In the forest, in the same world but with all of its trees, plants, roots, soil, and worms, the botanist could not calmly arrange the pieces of her jigsaw puzzle on her card table. Scattered through time and space, these leaves would never have met without her redistributing their traits into new combinations.

At the card table, with so many trumps in hand, every scientist becomes a structuralist. No need to look any further for the martingale that wins every time against those who sweat in the forest, those crushed beneath the complex phenomena that are maddeningly present, indiscernible, impossible to identify, reshuffle, and control. In losing the forest, we win knowledge of it. In a beautiful contradiction, the English word "oversight" exactly captures the two meanings of this domination by sight, since it means at once looking at something from above and ignoring it.

In the naturalist's collection things happen to plants that have never occurred since the dawn of the world (see Chapter 5). The plants find themselves detached, separated, preserved, classified, and tagged. They are then reassembled, reunited, redistributed according to entirely new principles that depend on the researcher, on the discipline of botany, which has been standardized for centuries, and on the institution that shelters them, but they no longer grow as they did in the great forest. The botanist learns new things, and she is transformed accordingly, but the plants are transformed also. From this point of view there is no difference between observation and experience: both are constructions. Through its displacement onto this table, the interface between forest and savanna becomes a hybrid mixture of scientist, botany, and forest, the proportions of which I will have to calculate later.

Still, the naturalist does not always succeed. In the upper-right-hand corner of the photograph something scary is brewing: an enormous pile of newspaper stuffed with plants brought back from the site and awaiting classification. The botanist has fallen behind. It is the same story in every laboratory. As soon as we go into the field or turn on an instrument, we find ourselves drowning in a sea of data. (I too have this problem, being incapable of saying all that can be said about a field trip that took only fifteen days.) Darwin moved out of his house soon after his voyage, pursued by treasure chests of data that ceaselessly arrived from the *Beagle*. Within the botanist's collection, the forest, reduced to its simplest expression, can quickly become as thick as the tangle of branches from which we started. The world can return to confusion at any point along this displacement: in the pile of leaves to be indexed, in the botanist's notes which threaten to submerge her, in the reprints sent from colleagues, in the library where the issues of journals are piling up. We have barely arrived when we must leave; the first instrument is hardly operational when we must think of a second device to absorb what its predecessor has already inscribed. The pace must be accelerated if we are to avoid being overwhelmed by worlds of trees, plants, leaves, paper, texts. Knowledge derives from such *movements*, not from simple contemplation of the forest.

We now know the advantages of being in an air-conditioned museum, but we have gone too quickly over the transformations that Edileusa made the forest undergo. I have opposed too abruptly the image of the botanist pointing to the trees and that of the naturalist in control of specimens on the worktable. In passing directly from the field to the collection, I must have missed the decisive go-between. If I say that "the cat is on the mat," I may seem to be designating a cat whose actual presence on said mat would validate my statement. In actual practice, however, one never travels directly from objects to words, from the referent to the sign, but always through a risky intermediary pathway. What is no longer visible with cats and mats, because they are too familiar, becomes visible again as soon as I take a more unusual and complicated statement. If I say "the forest of Boa Vista advances on the savanna" how can I point to that whose presence would accord a truth-value to my sentence? How can one engage those sorts of objects into discourse; to use an old word, how can one "educe" them into discourse? One needs to go back to the field and carefully follow, not only what happens inside collections, but how our friends are collecting data in the forest itself.

In the photograph in Figure 2.7, everything is a blur. We have left the laboratory and are now in the midst of the virgin forest. The researchers can only be distinguished as khaki and blue spots on a green background, and at any moment they could disappear into the Green Hell of the forest if they move away from one another.

René, Armand, and Héloïsa are having a discussion around a hole in the ground. Holes and pits are to pedology what a specimen collection is to botany: the basic craft and the focus of obsessive attention. Since the structure of soil is always hidden beneath our feet, pedologists can display its profile only by digging holes. A profile is the assemblage of the successive layers of soil, designated by the beautiful word "horizon." Rainwater, plants, roots, worms, moles, and billions of bacteria transform the parent material of the bedrock (studied by geologists) into many different "horizons," which the pedologists learn to distinguish, classify, and envelop in a history that they call "pedogenesis" (Ruellan and Dosso 1993).

In accordance with the habits of their profession, the pedologists wanted to know whether the bedrock was, at a certain depth, different beneath the forest than beneath the savanna. Here was a simple hypothesis that would have put an end to the controversy between botany and pedology: neither the forest nor the savanna is receding, the border that separates them reflects a difference in soil. The superstruc-

Figure 2.7

ture would be explained by the infrastructure, to use an old Marxist metaphor. Yet, as they soon discover, at depths below fifty centimeters the soil under the savanna and the soil under the forest appear exactly the same. The hypothesis from infrastructure does not hold. Nothing in the bedrock seems to explain the difference in the superficial horizons—clayey beneath the forest and sandy beneath the savanna. The profile is "bizarre," and that makes my friends all the more excited.

In the picture in Figure 2.8, René is standing and aiming at me with an instrument combining compass and clisimeter in order to establish a first topographic bearing. While taking advantage of the situation to snap a picture, I play the minor role, well suited to my height, of an alignment pole so that René can mark precisely where the pedologists should dig their holes. Lost in the forest, the researchers rely on one of the oldest and most primitive techniques for organizing space, claiming a place with stakes driven into the ground to delineate geometric shapes against the background noise, or at least to permit the possibility of their recognition.

Submerged in the forest again, they are forced to count on the oldest

Figure 2.8

of the sciences, the measure of angles, a geometry whose mythical origin has been recounted by Michel Serres (Serres 1993). Once more a science, pedology, must follow the tracks of an older discipline, surveying, without which we would dig our holes haphazardly, trusting to luck, incapable of creating on graph paper the precise map that René would like to draw. The succession of triangles will be used as a reference and will be added to the numbering of square sections of the field site already done by Edileusa (see Figure 2.3). In order for the botanical and pedological data to be superposed on the same diagram later, these two bodies of reference must be compatible. One should never speak of "data"—what is given—but rather of *sublata,* that is, of "achievements."

René's standard practice is to reconstitute the surface soil along transects, the extreme limits of which contain soils that are as different as possible. Here, for example, it is very sandy beneath the savanna and very clayey beneath the forest. He proceeds by approximate gradations, first choosing two extreme soils, then taking a sample in the middle. Starting again, he continues in this way until he obtains homogeneous horizons. His method recalls both artillery (it approximates by finding medians), and anatomy (it traces the geometry of horizons, true "organs" of the soil). If I were playing the historian, not the philosopher in pursuit of reference, I would discuss at length the fascinating paradigm of what René calls "structural pedology," how it distinguishes itself from others and the controversies that arise from it.

To get from one point to another the pedologists cannot use a surveyor's chain of measurement; no agriculturist has ever leveled this soil. Instead they use a wonderful instrument, the Topofil Chaix™ (Figure 2.9), a device that their Brazilian colleagues have perversely named a "pedofil," and of which Sandoval, in this photograph, reveals the mechanism by opening its orange box. So much depends upon an orange pedofil . . .

A spool of cotton thread unrolls evenly and spins a pulley that activates the cogwheel of a counter. Setting the counter to zero, then unwinding the thread of Ariadne behind him, the pedologist can get from one point to the next. Upon arrival at his destination, he simply cuts the thread with a blade set near the spool and ties off the end to prevent any untimely unrolling. A glance at the window on the counter tells the distance he has traveled to within a meter. His path becomes a single number easily transcribed into a notebook and—a double advantage—takes on material form in the thread that remains in place. Losing an expensive and distracted pedologist in the Green Hell is impossible: the cotton thread will always bring him back to camp. If Hansel and Gretel had had access to a "Topofil Chaix à fil perdu n° de référence I-8237," their tale would have unwound very differently.

After a few days' work the field site is littered with threads that entangle our feet. Still, as a result of the compass's measurements of angles and the pedofil's measurements of lines, the land has become a proto-laboratory—a Euclidean world where all phenomena can be registered by a collection of coordinates. Had Kant used this instrument, he would have recognized in it the practical form of his philosophy. For the world to become knowable, it must become a laboratory. If virgin forest is to be transformed into a laboratory, the forest must be prepared to be rendered as a diagram (Hirshauer 1991). In the extraction of a diagram from a confusion of plants, scattered locations become marked and measured points linked by cotton threads that materialize (or spiritualize) lines in a network composed of a succession

Figure 2.9

of triangles. Equipped only with the *a priori* forms of intuition, to use Kant's expression again, it would be impossible to draw these sites together, short of teaching, somehow, a limbless mind-in-a-vat how to use such equipment as compasses, clisimeters, and topofils.

Sandoval the technician, the only person on the expedition who is native to the region, has dug the largest part of the hole shown in Figure 2.10. (Of course had I not artificially severed the philosophy from the sociology, I would have to account for this division of labor between French and Brazilians, mestizos and Indians, and I would have to explain the male and female distributions of roles.) Armand, here leaning on the drill, is removing core samples by collecting earth in the small chamber at its tip. Unlike Sandoval's tool, the mattock that is lying on the ground now that its task is complete, the drill is a piece of laboratory equipment. Two rubber stoppers placed at 90 centimeters and at one meter allow it to be used both as an instrument for measuring depth and, by pushing and twisting, as a sampling tool. The pedologists examine the soil sample, then Héloïsa collects it in a plastic bag on which she writes the number of the hole and the depth at which it was taken.

Figure 2.10

As with Edileusa's specimens, most of the analyses cannot be performed in the field but must be done in the laboratory. The plastic bags here begin a long voyage that will take some of them to Paris, via Manaus and São Paulo. Even if René and Armand are able to judge on the spot the quality of the earth, its texture, its color, and the activity of earthworms, they cannot analyze the soil's chemical composition, its grain size, or the radioactivity of the carbon it contains without costly instruments and skill that one does not easily find among the poor *garimperos* or the wealthy landowners. On this expedition, the pedologists are the vanguard for the distant laboratories to which they will take their samples. The samples will remain attached to their original context solely by the fragile link of the numbers inscribed in black felt-tip pen on the little transparent bags. If, like me, you should ever run into a gang of pedologists, one word of advice: never offer to carry their suitcases, which are enormous and stuffed with the bags of earth they tote from one part of the world to another and with which they will quickly fill your refrigerator. The circulation of their samples traces a network on the Earth as dense as the cotton webs spun by their topofils.

What industrialists call the "traceability" of references depends, in this case, on the reliability of Héloïsa. Sitting in front of the hole, the group members rely on her for the careful maintenance of the field notebook. For each sample she must record the coordinates of the location, the number of the hole, the time and depths at which it was collected. In addition, she must note down all the qualitative data her two male colleagues can extract from the lumps of earth before they slide them into the bags.

The success of the entire expedition depends on this little logbook, equivalent to the protocol book that regulates the life of any laboratory. It is this book that will allow us to return to each data point in order to reconstitute its history. The list of questions that was decided on at the restaurant is imposed on each sequence of action by Héloïsa. It is a grid that we must systematically fill with information. Héloïsa acts as guarantor of the standardization of experimental protocols, so that we take the same kinds of samples from each location and in the same way. The protocols ensure the compatibility and therefore the comparability of the holes, and the notebook then allows for continuity in time as well as in space. Héloïsa does not only handle tags and protocols. A geomorphologist, she adds her two cents to all the conversations, allowing her expatriate colleagues to "triangulate" their judgments through hers.

Listening to Héloïsa call us to order—having repeated the information dictated to us by René and twice verified the inscriptions on the bags—it seems to me that never before has the forest of Boa Vista known such discipline. The indigenous people who once traveled through this place probably imposed rites on themselves as well, perhaps as fastidious as those of Héloïsa, but surely not so strange. Sent by institutions that are thousands of kilometers away, obliged at all costs to maintain the traceability of the data we produce with minimal deformation (while transforming them totally by ridding them of their local context), we would have seemed extremely exotic to the indigenous people. Why take such care in sampling specimens whose features are visible only at such a distance that the context from which they were taken will have disappeared? Why not remain in the forest? Why not "go native"? And what about me, standing here, useless, arms dangling, incapable of distinguishing a profile from a horizon— am I not even more exotic, exacting from the hard labor of my informants the bare minimum for a philosophy of reference that will be of interest only to a very few colleagues in Paris, California, or Texas? Why not become a pedologist? Why not become an indigenous soil collector, an autochthonous botanist?

To understand these small anthropological mysteries we must draw closer to the beautiful object in Figure 2.11, the "pedocomparator." On the savanna grass, we see a series of empty little cardboard cubes aligned to form a square. More Cartesian coordinates, more columns, more rows. These little cubes rest in a wooden frame that allows them to be stowed away in a drawer. With the cleverness of our pedologists, and with the addition of a handle, clasps, and a padded flap that serves as a flexible cover for all the cardboard cubes (not visible in the photograph), this drawer can also be transformed into a suitcase. The suitcase permits the simultaneous transportation of all the clods of earth that have since become Cartesian coordinates, and their collection in what thus becomes a pedolibrary.

Like the cabinet in Figure 2.5, the pedocomparator will help us grasp the *practical* difference between abstract and concrete, sign and furniture. With its handle, its wooden frame, its padding, and its card-

Figure 2.11

board, the pedocomparator belongs to "things." But in the regularity of its cubes, their disposition in columns and rows, their discrete character, and the possibility of freely substituting one column for another, the pedocomparator belongs to "signs." Or rather, it is through the cunning invention of this hybrid that the world of things may become a sign. With the next three photographs we will try to understand more concretely the practical task of abstraction and what it means to load a state of affairs into a statement.

I will be obliged to employ vague terms—we do not have as discriminating a vocabulary for speaking of the engagement of things into discourse as we do for speaking of discourse itself. Analytic philosophers keep themselves busy trying to discover how we can speak of the world in a language capable of truth (Moore 1993). Curiously, even though they attach importance to the structure, coherence, and validity of language, in all their demonstrations the world simply awaits designation by words whose truth or falsehood is guaranteed solely by its presence. The "real" cat waits quietly on its proverbial mat to confer a truth-value on the sentence "the cat is on the mat." Yet to achieve certainty the world needs to stir and transform *itself* much more than *words* (see Chapters 4 and 5). It is this, the other neglected half of analytic philosophy, that analysts must now acknowledge.

For the time being, the pedocomparator is empty. This instrument can be added to the list of empty forms that has been getting longer during the expedition: Edileusa's plot of land, divided into squares by numbers inscribed on tags that are nailed to trees; the marking of the holes with René's compass and topofil; the numbering of the samples and the disciplined sequence of the protocol controlled by Héloïsa. All these empty forms are set up *behind* the phenomena, *before* the phenomena manifest themselves, *in order* for them to be manifested. Obscured in the forest by their sheer number, phenomena will be able at last to appear, that is, to stand out against the new backgrounds we have astutely placed behind them. In my eyes and in those of my friends, pertinent traits will be bathed in a spotlight as white as the empty pedocomparator or the graph paper, very different in any case from the deep greens and grays of the vast and noisy forest, where some birds whistle so obscenely that the locals call them "flirting birds."

In Figure 2.12, René abstracts. After cutting the earth with a knife, he removes a clod, from a depth dictated by the protocol, and deposits it in one of the cardboard cubes. With a felt-tip pen Héloïsa will code the edge of the cube with a number that she will also record in her notebook.

Consider this lump of earth. Grasped by René's right hand, it retains all the materiality of soil—"ashes to ashes, dust to dust." Yet as it is placed inside the cardboard cube in René's left hand, the earth becomes a sign, takes on a geometrical form, becomes the carrier of a numbered code, and will soon be defined by a color. In the philosophy of science, which studies only the resulting abstraction, the left hand does not know what the right hand is doing! In science studies, we are ambidextrous: we focus the reader's attention on this hybrid, this moment of substitution, the very instant when the future sign is abstracted from the soil. We should never take our eyes off the material weight of this action. The earthly dimension of Platonism is revealed in this image. We are not jumping from soil to the Idea of soil, but from continuous and multiple clumps of earth to a discrete color in a geometric cube coded in x- and y-coordinates. And yet René does not *impose* predetermined categories on a shapeless horizon; he *loads* his

Figure 2.12

pedocomparator with the meaning of the piece of earth—he educes it, he articulates* it (see Chapter 4). Only the movement of substitution by which the real soil becomes the soil known to pedology counts. The immense abyss separating things and words can be found everywhere, distributed to many smaller gaps between the clods of earth and the cubes-cases-codes of the pedocomparator.

What a transformation, what a movement, what a deformation, what an invention, what a discovery! In jumping from the soil to the drawer, the piece of earth benefits from a means of transportation that no longer transforms it. In the previous photograph we could see how the soil changed states; in Figure 2.13 we see how it changes location. Having made the passage from a clump of earth to a sign, the soil is now able to travel through space without further alterations and to remain intact through time. At night, in the restaurant, René opens the cabinet-suitcases of the two pedocomparators and contemplates the series of cardboard cubes regrouped in rows corresponding to holes and columns corresponding to depths. The restaurant becomes the annex of a pedolibrary. All the transects have become compatible and comparable.

Once filled, the cubes gather clods of earth on the way to becoming signs, but we know that the empty compartments, either humble ones like these or famous ones like those of Mendeleev, are always the most important part of any classification scheme (Bensaude-Vincent 1986; Goody 1977). When we compare them, the compartments define what is left for us to find, and we are able to plan the next day's labor in advance since we know what we must gather. Thanks to the empty compartments, we see the blanks in our protocol. According to René, "It is the pedocomparator that *tells us* if we have finished a transect."

The first great advantage of the pedocomparator, as "profitable" as the botanist's classification in Figure 2.6, is that in it all the different samples from all the different depths become visible simultaneously, though they were extracted over the course of a week. Thanks to the pedocomparator, the differences in color become manifest and form a table or chart; all of the disparate samples are embraced synoptically. The forest-savanna transition has now been translated, through the arrangement of nuanced shades of brown and beige, into columns and rows—a transition now graspable because the instrument has given us a handle on the earth.

Figure 2.13

Look at René in the photograph: he is master of the phenomenon that a few days earlier was tucked away in the soil, invisible, and dispersed in an undifferentiated continuum. I have never followed a science, rich or poor, hard or soft, hot or cold, whose moment of truth was not found on a one- or two-meter-square flat surface that a researcher with pen in hand could carefully inspect (see Figures 2.2 and 2.6). The pedocomparator has made the forest-savanna transition into a laboratory phenomenon almost as two-dimensional as a diagram, as readily observed as a map, as easily reshuffled as a pack of cards, as simply transported as a suitcase, about which René jots down notes while peacefully smoking his pipe, having taken a shower to wash off the dust and earth that are no longer useful.

And I, of course, ill-equipped and thus short on rigor, I bring back to the reader, by superposing pictures and text, a phenomenon, that of the *circulating reference**, that was until now invisible, purposely muddled by epistemologists, dispersed in the practice of scientists, and sealed up in the knowledges that I now calmly display with a cup of tea in hand at my house in Paris, while reporting what I observed at the border of Boa Vista.

Another advantage of the pedocomparator, once it is saturated with data: a pattern emerges. And here again, as with Edileusa's discoveries, it would be astounding were this not the case. Invention almost always follows the new handle offered by a new translation or transportation. The most incomprehensible thing in the world would be for the pattern to remain incomprehensible after such rearrangements.

This expedition, it too, via the intermediary of the pedocomparator, discovers or constructs (we will choose between those two verbs in Chapter 4, before realizing in Chapter 9 why we do not have to choose) an extraordinary phenomenon. Between the sandy savanna and the clayey forest, it seems that a twenty-meter-wide strip of land spreads out at the border, on the savanna side. This strip of land is ambiguous, more clayey than the savanna but less so than the forest. It would appear that the forest casts its own soil before it to create conditions favorable to its expansion. Unless, on the contrary, the savanna is degrading the woodland humus as it prepares to invade the forest. The various scenarios that my friends discuss, at night in the restaurant, are now gauged by the weight of evidence. They become possible interpretations of the matters of fact that are solidly in place in the grid of the pedocomparator.

One scenario will eventually become text, and the pedocomparator will become a table in an article. There now needs to be only one last, tiny transformation.

On the table, in the table/chart, in Figure 2.14, we see the forest on the left and the savanna on the right, the reverse of Figure 2.1, give or take a few transformations. (Since there are not enough compartments in the pedocomparator, the series of samples must be altered, breaking the beautiful order of the table and requiring us to devise an ad hoc reading convention.) Beside the open drawers there is a diagram drawn on millimeter-ruled graph paper and a table drawn on straight-ruled paper. The coordinates of the samples, taken by the team along a given transect, are recaptured in a vertical cross-section, while the chart sums up color variations as a function of depth at a given set of coordinates. A transparent ruler negligently placed on the drawer further ensures the transition from furniture to paper.

In Figure 2.12 René moved from concrete to abstract in one quick gesture. He was moving from thing to sign and from the three-dimensional earth to the two and a half dimensions of the table/chart. In Figure 2.13 he had slipped from the field site to the restaurant: the drawers convert into a suitcase, permitting René's movement from an uncomfortable and underequipped location to the relative comfort of a café, and in principle nothing (except Customs officers) can stop the transportation of this drawer/suitcase/chart anywhere in the world, or its comparison with all other profiles in all other pedolibraries.

In Figure 2.14 another transformation as important as the others becomes evident, but one that, under the name of inscription*, has received more attention than the others. We move now from the instrument to the diagram, from the hybrid earth/sign/drawer to paper.

People are often surprised that mathematics can be applied to the world. In this case, for once, the surprise is misplaced. For here we must ask how much the world needs to change in order for one kind of paper to be *superposed* on a geometry of another kind without suffering too much distortion. Mathematics has never crossed the great abyss between ideas and things, but it is able to cross the tiny gap between the already geometrical pedocomparator and the piece of millimeter-ruled paper on which René has recorded the data from the sam-

Figure 2.14

ples. It is easy to cross this gap—I can even measure the distance with a plastic ruler: ten centimeters!

As abstract as the pedocomparator is, it remains an object. It is lighter than the forest, yet heavier than the paper; it is less corruptible than the vibrant earth, but more corruptible than geometry; it is more mobile than the savanna, but less mobile than the diagram that I could send by phone if Boa Vista had a fax machine. As coded as the pedocomparator is, René cannot insert it into the text of his report. He can only hold it in reserve, keeping it for future comparisons if he ever begins to have doubts about his article. With the diagram, in contrast, the forest-savanna transition becomes paper, assimilable by every article in the world, and transportable to every text. The geometric form of the diagram renders it compatible with all the geometric transformations that have ever been recorded since *centers of calculation** have existed. What we lose in matter through successive reductions of the soil, we regain a hundredfold in the branching off to other forms that such reductions—written, calculated, and archival—make possible.

In the report that we are preparing to write, only one rupture will remain, a gap as tiny and as immense as all the steps we have just followed: I mean the gap that divides our prose from the annex of diagrams it will refer to. We will write about the forest-savanna transition, which we will show within the text through the medium of a graph. The scientific text is different from all other forms of narrative. It speaks of a referent, *present* in the text, in a form other than prose: a chart, diagram, equation, map, or sketch. Mobilizing its own *internal* referent*, the scientific text carries within itself its own verification.

In Figure 2.15 is the diagram that combines all the data obtained during the expedition. It appears as "figure 3" in the written report of which I am one of the proud authors and of which the title page reads:

Relations between Vegetation Dynamics and the Differentiation of Soils in the Forest-Savanna Transition Zone in the Region of Boa Vista, Roraima, Amazonia (Brazil) Report on Expedition in Roraima Province, October 2–
14, 1991
E. L. Setta Silva (1), R. Boulet (2), H. Filizola (3),
S. do N. Morais (4), A. Chauvel (5) and B. Latour (6)
(1) MIRR, Boa Vista RR, (2.3) USP, São Paulo, (3-5)
INPA,
Manaus, (6) CSI, ENSMP, (2.5) ORSTOM Brazil

Let us quickly retrace our steps back down the road we have traveled while following our friends. The prose of the final report speaks of a diagram, which summarizes the form displayed by the layout of the pedocomparator, which extracts, classifies, and codes the soil, which, in the end, is marked, ruled, and designated through the crisscrossing of coordinates. Notice that, at every stage, each element belongs to matter by its origin and to form by its destination; it is abstracted from a too-concrete domain before it becomes, at the next stage, too concrete again. We never detect the rupture between things and signs, and we never face the imposition of arbitrary and discrete signs on shapeless and continuous matter. We see only an unbroken series of well-nested elements, each of which plays the role of sign for the previous one and of thing for the succeeding one.

At every stage we find elementary *forms* of mathematics, which are used to collect *matter* through the mediation of a practice embodied

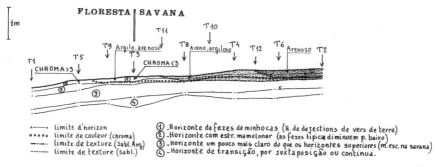

Figure 3. Coupe du transect 1

Figure 2.15

in a group of researchers. On each occasion a new phenomenon is educed from this hybrid of form, matter, skilled bodies, and groups. Let us remember René, in Figure 2.12, placing the brown earth into the white cardboard cube that was then immediately marked with a number. He did not divide the soil according to intellectual categories, as in the Kantian mythology; rather, he conveyed the meaning of each phenomenon by making matter cross the gap that separated it from form.

In fact, if we flip quickly through these photographs, we become aware that, even if my inquiry had been more meticulous, each stage would reveal a rift as complete as those which follow and precede it. Try as I might, like a new Zeno, to multiply the intermediaries, there is never a *resemblance* between stages so that we can merely superpose them. Compare the two extremes in Figures 2.1 and 2.15. The difference between them is no wider than that between the lumps of earth sampled by René (Figure 2.12) and the data-points that they become in the pedocomparator. Whether I choose the two extremes or multiply the intermediaries, I find this same discontinuity.

Yet there is also a continuity, since all the photographs say the same thing and represent the same forest-savanna transition, made ever more certain and precise at each stage. Our field report indeed refers to "figure 3," which indeed refers to the Boa Vista forest. Our report refers to the strange dynamics of vegetation that appear to allow the forest to defeat the savanna, as if the trees had turned sandy soil into clay to prepare for growth in the twenty-meter-wide strip of land. But these acts of reference are all the more assured since they rely not so much on resemblance as on a regulated series of transformations, transmutations, and translations. A thing can remain more durable and be transported farther and more quickly if it continues to undergo transformations at each stage of this long cascade.

It seems that reference is not simply the act of pointing or a way of keeping, on the outside, some material guarantee for the truth of a statement; rather it is our way of keeping something *constant* through a series of transformations. Knowledge does not reflect a real external world that it resembles via mimesis, but rather a real interior world, the coherence and continuity of which it helps to ensure. What a beautiful move, apparently sacrificing resemblance at each stage only to settle again on the same meaning, which remains intact through sets of rapid transformations. The discovery of this strange and contradictory behavior is worthy of the discovery of a forest able to create its own soil. If I could find the solution to that puzzle, my own expedition would be no less productive than that of my happy colleagues.

In order to understand the constant that is maintained throughout these transformations, let us consider a small apparatus as ingenious as the topofil or the pedocomparator (Figure 2.16). Since our friends cannot easily bring the soil of Amazonia back to France, they must be able to transform the color of each cube using a label, and if possible a number, that will make the samples of soil compatible with the universe of calculation and allow the scientists to benefit from the advantage that all calculators lend to every manipulator of signs.

But won't relativism rear its monstrous head as we attempt to qualify the nuances of brown? How can we dispute tastes and colors? As the French saying goes, "So many heads, so many opinions." In Figure 2.16 we see René's solution for repairing the ravages of relativism.

For thirty years he has toiled in the tropical soils of the world carrying a small notebook with rigid pages: the Munsell code. Each page of this little volume groups together colors of very similar shades. There is a page for the purplish reds, another for the yellowish reds, another for the browns. The Munsell code is a relatively universalized norm; it is used as a common standard for painters, paint manufacturers, cartographers, and pedologists, since page by page it arranges all the nuances of all the colors of the spectrum by assigning each a number.

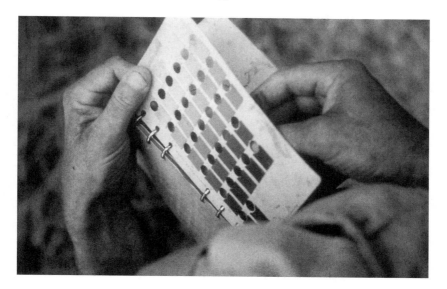

Figure 2.16

The number is a reference that is quickly understandable and repro-
ducible by all the colorists in the world on the condition that they use
the same compilation, the same code. By telephone, you and a salesperson cannot match samples of wallpaper, but you can, based on a
color chart the salesperson has given you, select a reference number.

The Munsell code is a decisive advantage for René. Lost in Roraima,
made so tragically local, he is able to become, through the intermediary of his code, as global as it is possible for a human being to be. The
unique color of this particular soil sample becomes a (relatively) universal number.

At this moment, the power of standardization (Schaffer 1991) is less
interesting to me than a stupefying technical trick—the little holes
that have been pierced above the shades of color. Though seemingly
always out of reach, the threshold between local and global can now
be crossed instantaneously. Still, it takes some skill to insert the soil
sample into the Munsell code. In order for the soil sample to qualify as
a number, René must in fact be able to match, superpose, and align the
local clod of earth, which he holds in his hand, with the standardized
color chosen as a reference. To accomplish this, he passes the soil sam-

ple beneath the openings made in the notebook and, by successive approximations, selects the color closest to that of the sample.

There is, as I have said, a complete rupture at each stage between the "thing" part of each object and its "sign" part, between the tail end of the soil sample and its head. That abyss is all the wider because our brains are incapable of memorizing color with precision. Even if the soil sample and the standard were no farther apart than ten or fifteen centimeters, the width of the notebook, this would be enough for René's brain to forget the precise correspondence between the two. The only way the resemblance between a standardized color and a soil sample can be established is by piercing holes in the pages that allow us to align the rough surface of the lump of soil with the bright and uniform surface of the standard. With less than a millimeter of distance separating them, then and only then can they be read synoptically. Without the holes, there can be no alignment, no precision, no reading, and therefore no transmutation of local earth into universal code. Across the abyss of matter and form, René throws a bridge. It is a footbridge, a line, a grappling hook.

"The Japanese have made one without holes," René says; "I cannot use it." We are always amazed by the minds of scientists, and justly so, but we should also admire their utter lack of trust in their own cognitive abilities (Hutchins 1995). They doubt their brains so much that they need to invent little tricks like this to ensure their understanding of the simple color of a soil sample. (And how could I make the reader understand this work of reference without the photographs that I have taken, images that must be viewed at exactly the same time as the story I am relating is being read? I am so afraid of making a mistake in my account that I myself do not dare lose sight of the photographs, even for an instant.)

The rupture between the handful of dust and the printed number is always there, though it has become infinitesimal because of the holes. Through the intermediary of the Munsell code, a soil sample can be read as a text: "10YR3/2"—further evidence of the practical Platonism that turns dust into an Idea via the two callused hands firmly holding a notebook/instrument/calibrator.

Let us follow in more detail the trail displayed in Figure 2.16, sketching the lost road of reference for ourselves. René has extracted his lump of earth, renouncing the too rich and too complex soil. The hole, in turn, allows the framing of the lump and the selection of its color by ignoring its volume and texture. The little flat rectangle of color is then used as an intermediary between the earth, summarized as a color, and the number inscribed under the corresponding shade. Just as we are able to ignore the volume of the sample in order to concentrate on the color of the rectangle, we are soon able to ignore the color in order to conserve only the reference number. Later, in the report, we will omit the number, which is too concrete, too detailed, too precise, and retain only the horizon, the tendency.

Here we find the same cascade as before, of which only a tiny portion (the passage from the sample color to that of the standard) rests on resemblance, on *adequatio*. All the others depend only on the conservation of traces that establish a reversible route that makes it possible to retrace one's footsteps as needed. Across the variations of matters/forms, scientists forge a pathway. Reduction, compression, marking, continuity, reversibility, standardization, compatibility with text and numbers—all these count infinitely more than *adequatio* alone. No step—except one—resembles the one that precedes it, yet in the end, when I read the field report, I am indeed holding in my hands the forest of Boa Vista. A text truly speaks of the world. How can resemblance result from this rarely described series of exotic and minuscule transformations obsessively nested into one another so as to keep something constant?

In Figure 2.17 we see Sandoval squatting, the shaft of the mattock still resting under his arm, contemplating the new hole he has just dug. Standing, Héloïsa is thinking about the few animals in this green-gray forest. She is wearing a geologist's pouch, an ammunition belt the side of which is studded with eyelets too narrow for cartridges but well suited for carrying the colored pencils indispensable to the professional cartographer. In her hand she holds the famous notebook, the protocol book that makes it true that we are in a vast, green laboratory. She is waiting to open it and to take notes now that both pedologists have finished their examination and reached agreement.

Armand (on the left) and René (on the right) are engaged in the rather strange exercise of "earth tasting." In one hand each of them has taken a bit of soil sampled from the hole at a depth dictated by Héloïsa's protocol. They have delicately spat on the dust and now, with the other hand, they slowly knead it. Is this for the pleasure of

Figure 2.17

molding figurines? No, it is to extract another judgment, one that no longer involves color, but rather texture. Unfortunately, for this purpose there is no equivalent of the Munsell code, and if there were one, we wouldn't know how to get it here. To define granularity in a standardized manner, one would need half of a well-equipped laboratory. Consequently, our friends must content themselves with a qualitative test that rests on thirty years' experience and that they will later compare with laboratory results. If the soil is easily molded, it is clay; if it crumbles under one's fingers, then one is dealing with sand. Here is an apparently very easy trial that amounts to a sort of laboratory experiment in the hollow of one's hand. The two extremes are easily recognizable, even by a beginner like me. It is the intermediate compounds of sand and clay that make the differentiation difficult and crucial, since we are interested in qualifying the subtle modifications of the transition soils which are more clayey toward the forest and more sandy toward the savanna.

Lacking any kind of gauge, Armand and René rely on a back-and-forth discussion of their judgments of taste, as my father would do when he tasted his Corton wines.

"Sandy-clay or clayey-sand?"

"No, I would say clayey, sandy, no sandy-clay."

"Wait, mold it a bit more, give it some time."

"Okay, yes, let's say between sandy-clay and clayey-sand."

"Héloïsa, make a note: at P2, between five and seventeen centimeters, *areno-argiloso a argilo-arenoso.*" (I forgot to mention that we are alternating constantly between French and Portuguese, the politics of language being added to the politics of race, gender, and disciplines.)

The combination of discussion, know-how, and physical manipulation allows for the extraction of a calibrated qualification of texture that can immediately replace, in the notebook, the soil that can now be thrown away. A word replaces a thing while conserving a trait that defines it. Is this a term-to-term correspondence? No, the judgment does not *resemble* the soil. Is this metaphorical displacement? No more so than a correspondence. Is it metonymy? Not that either, since once we take a handful of soil for the whole horizon, we keep only what is on the paper of the notebook and none of the earth that was used to qualify it. Is this compression of data? Yes, definitely, since four words occupy the location of the soil sample, but it is a change of state so radical that now a sign appears in place of a thing. Here it is no longer a question of reduction but of transubstantiation.

Are we crossing the sacred boundary that divides the world from discourse? Obviously yes, but we have already crossed it a good ten times. This new leap is no more distant than the preceding one, in which the earth extracted by René, cleaned of blades of grass and worm feces, became evidence in a test of its resistance to molding; or the one before that, in which Sandoval dug the P2 hole with his mattock; or the following one, in which, on the diagram, the whole horizon from five to seventeen centimeters takes on a single texture, allowing, through induction, the coverage of the surface from a point; or the n+1 transformation that permits a diagram drawn on millimeter-ruled graph paper to play the role of internal referent for the written report. There is nothing privileged about the passage to words, and all stages can serve equally to allow us to grasp the nesting of reference. In none of the stages is it ever a question of copying the preceding stage. Rather, it is a matter of *aligning* each stage with the ones that precede and follow it, so that, beginning with the last stage, one will be able to *return* to the first.

How can we qualify this relation of representation, of delegation, when it is not mimetic yet is so regulated, so exact, so packed with reality, and, in the end, so realistic? Philosophers fool themselves when they look for a correspondence between words and things as the ultimate standard of truth. There is truth and there is reality, but there is neither correspondence nor *adequatio*. To attest to and guarantee what we say, there is a much more reliable movement—indirect, crosswise, and crablike—through successive layers of transformations (James [1907] 1975). At each step, most of the elements are lost but also renewed, thus leaping across the straits that separate matter and form, without aid other than, occasionally, a resemblance that is more tenuous than the rails that help climbers over the most acrobatic passes.

In Figure 2.18 we are on the site, toward the end of our expedition, and René is commenting on a diagram on graph paper of a vertical cross-section of the transect that we have just dug and examined. Torn, dirty, stained with sweat, incomplete, and sketched in pencil, this diagram is the direct predecessor of the one in Figure 2.15. From the one to the other there are indeed transformations, which include processes of selection, centering, lettering, and cleaning, but these

Figure 2.18

are minor in comparison with the transformations through which we have just passed (Tufte 1984).

In the middle of the photograph René is indicating a line with his finger, a gesture we have followed from the first (see Figures 2.1 and 2.2). Unless it is pointed in anger as a prelude to a fist, the extension of the index finger always signals an access to reality even when it targets a mere piece of paper, an access which in this case nonetheless encompasses the totality of the site, which, paradoxically, has entirely disappeared even as we are sweating at the center of it. This is the same reversal of space and time we have already seen many times: thanks to inscriptions, we are able to oversee and control a situation in which we are submerged, we become superior to that which is greater than us, and we are able to gather together synoptically all the actions that occurred over many days and that we have since forgotten.

But the diagram not only redistributes the temporal flux and inverts the hierarchical order of space, it reveals to us features that previously were invisible even though they were literally under the feet of our pedologists. It is impossible for us to see the forest-savanna transition in vertical cross-sections, to qualify it in homogeneous horizons, and to mark it with data-points and lines. René points with his finger made of flesh and attracts the gaze of the living onto a profile whose observer could never exist. The observer would have to not only reside under the earth like a mole but be able to cut the soil as if with a blade hundreds of meters long and replace the confusing variation of forms with homogeneous hatchings! To say that a scientist "occupies a standpoint" is never very useful, since she will immediately move to another through the application of an instrument. Scientists never *stand* in their standpoint.

Despite the implausible vista it offers, the diagram adds to our information. On one paper surface we combine very different sources that are blended through the intermediary of a homogeneous graphical language. The positions of the samples along the transect, the depths, the horizons, the textures, and the reference numbers of the colors can be added to one another by superposition—and the reality we had lost is replaced.

René, for instance, has just added to the diagrams the worm feces I have mentioned. According to my friends, it seems that the worms may carry the solution to the enigma within their particularly voracious digestive tracts. What produces the strip of clayey soil in the savanna at the edge of the forest? Not the forest, since this strip extends twenty meters beyond the protective shadow and nourishing humidity of the trees. Not the savanna either, since, let us remember, it always reduces clay into sand. What is this mysterious action at a distance that prepares the soil for the arrival of the forest, ascending the thermodynamic slope that continues to degrade the clay? Why not the earthworms? Might they be the catalyzing agents of the pedogenesis? In modeling the situation, the diagram allows for the imagining of new scenarios, which our friends discuss passionately while considering what is missing and where to dig the next hole to get back to the "raw data" with their pick and drill (Ochs, Jacoby, et al. 1994).

Is the diagram that René holds in his hand more abstract or more concrete than our previous stages? More abstract, since here an infinitesimal fraction of the original situation is preserved; more concrete, since we can grasp in our hands, and see with our eyes, the essence of the forest-savanna transition, summarized in a few lines. Is the diagram a construction, a discovery, an invention, or a convention? All four, as always. The diagram is *constructed* by the labors of

five people and by passing through successive geometrical constructions. We are well aware that we have *invented* it and that, without us
and the pedologists, it would never have appeared. Still, it *discovers* a
form that until now has been hidden but that we retrospectively feel
was already there beneath the visible features of the soil. At the same
time, we know that without the *conventional* coding of judgments,
forms, tags, and words, all we could see in this diagram drawn from
the earth would be formless scribbles.

All of these contradictory qualities—contradictory, that is, for us
philosophers—ballast this diagram with reality. It is not realistic; it
does not resemble anything. It does *more* than resemble. It *takes the
place of the original situation,* which we can retrace, thanks to the protocol book, the tags, the pedocomparator, the record cards, the stakes,
and, finally, the delicate spiderweb woven by the "pedofil." Yet we
cannot divorce this diagram from this series of transformations. In
isolation, it would have no further meaning. It replaces without replacing anything. It summarizes without being able to substitute completely for what it has gathered. It is a strange transversal object, an
alignment operator, truthful only on condition that it allow for *passage*
between what precedes and what follows it.

On the last day of the expedition we find ourselves in the restaurant,
now transformed into a meeting room for our mobile laboratory, in
order to write a draft of our report (Figure 2.19). René is holding the
now completed diagram in his hand and commenting on it, pointing with a pencil for the benefit of Edileusa and Héloïsa. Armand has
just finished reading the only thesis that has been published on our
corner of the forest, and he has opened it to pages of color photographs obtained by satellite. In the foreground rest the notebooks of
the anthropologist who is taking this picture—one more form of re-
cording amid forms of inscription. We are again among maps and
signs, two-dimensional documents and published literature, already
quite far from the site where we have labored for ten days. Have we,
then, returned to our starting point (see Figure 2.2)? No, because we
now have *gained* these diagrams, these new inscriptions we are attempting to interpret and to insert as an appendix and as evidence
into a narrative we are negotiating together, paragraph by paragraph,
in two languages, French and Portuguese. Let me quote a passage from
page one:

Figure 2.19

The interest of this expedition report stems from the fact that, in the first phase of work, the conclusions of the approaches of botany and pedology appear contradictory. *Without the contribution of the botanical data, the pedologists would have concluded that the savanna is advancing on the forest.* The collaboration of the two disciplines in this case has forced us to ask new questions of pedology. (italics in the original)

Here we are on much more familiar terrain—rhetoric, discourse, epistemology, and the writing of articles—busy with the weighing of arguments for and against the advance of the forest. Neither philosophers of language, nor sociologists of controversy, nor semioticians, nor rhetoricians, nor scholars of literature will have much difficulty here.

As thrilling as will be the transformations that Boa Vista will undergo from text to text, I do not, for the moment, wish to follow them. What interests me now is the transformation undergone by the soil, now bound up in words. How to summarize this? I need to draw, not a diagram on graph paper like that of my colleagues, but at least a sketch, a schema that will allow me to locate and point to what I, in my own field of science studies, have discovered: a discovery brought back from the underworld, worthy of our lowly brethren, the earthworms.

The philosophy of language makes it seem as if there exist two disjointed spheres separated by a unique and radical gap that must be reduced through the search for correspondence, for reference, between words and the world (Figure 2.20). While following the expedition to Boa Vista, I arrived at a quite different solution (Figure 2.21). Knowledge, it seems, does not reside in the face-to-face confrontation of a mind with an object, any more than reference designates a thing by means of a sentence verified by that thing. On the contrary, at every stage we have recognized a common operator, which belongs to matter at one end, to form at the other, and which is separated from the stage that follows it by a gap that no resemblance could fill. The operators are linked in a series that *passes across* the difference between things and words, and that redistributes these two obsolete fixtures of the philosophy of language: the earth becomes a cardboard cube, words become paper, colors become numbers, and so forth.

An essential property of this chain is that it must remain *reversible*. The succession of stages must be traceable, allowing for travel in both directions. If the chain is interrupted at any point, it ceases to transport truth—ceases, that is, to produce, to construct, to trace, and to conduct it. *The word "reference" designates the quality of the chain in its entirety,* and no longer *adequatio rei et intellectus.* Truth-value *circulates* here like electricity through a wire, so long as this circuit is not interrupted.

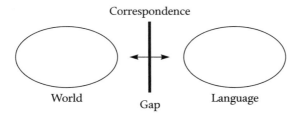

Figure 2.20 The "saltationist's" (James [1907] 1975) conception of the feat of correspondence implies that there is a gap between world and words that reference aims to bridge.

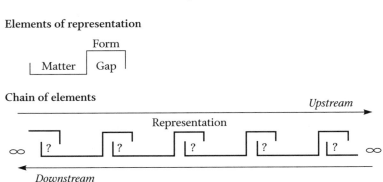

Figure 2.21 The "deambulatory" conception of reference follows a series of transformations, each of them implying a small gap between "form" and "matter"; reference, in this view, qualifies the movement back and forth as well as the quality of
the transformation; the key point is that reference, in this model, grows from the
center toward the two extremities.

Another property is revealed by the comparison of my two
sketches: the chain has no limit at either end. In the prior model (Figure 2.20), the world and language existed as two finite spheres capable
of self-enclosure. Here, on the contrary, we can elongate the chain
indefinitely by extending it at both ends, by adding other stages—yet
we can neither cut the line nor skip a sequence, despite our capacity to
summarize them all in a single "black box."

In order to understand the chain of transformation, and to grasp the
dialectic of gain and loss that, as we have seen, characterizes each
stage, we must look from above as well as at the cross-section (Figure
2.22). From forest to expedition report, we have consistently re-
represented the forest-savanna transition as if drawing two isosceles
triangles covering each other in reverse. Stage by stage, we lost locality, particularity, materiality, multiplicity, and continuity, such that,
in the end, there was scarcely anything left but a few leaves of paper.
Let us give the name *reduction* to the first triangle, whose tip is all
that finally counts. But at each stage we have not only reduced, we
have also gained or regained, since, with the same work of re-
representation, we have been able to obtain much greater compatibility, standardization, text, calculation, circulation, and relative universality, such that by the end, inside the field report, we hold not only

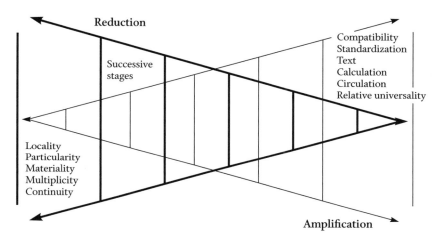

Figure 2.22 The transformation at each step of the reference (see Figure 2.21) may be pictured as a trade-off between what is gained (amplification) and what is lost (reduction) at each information-producing step.

all of Boa Vista (to which we can return), but also the explanation of its dynamic. We have been able, at every stage, to extend our link with already-established practical knowledge, starting with the old trigonometry placed "behind" phenomena and ending up with all of the new ecology, the new findings of "botanical pedology." Let us call this second triangle, by which the tiny transect of Boa Vista has been endowed with a vast and powerful basis, *amplification.*

Our philosophical tradition has been mistaken in wanting to make phenomena* the meeting point between things-in-themselves and categories of human understanding (Figure 2.23; also see Chapter 4). Realists, empiricists, idealists, and assorted rationalists have fought ceaselessly among themselves around this bipolar model. Phenomena, however, are not found at the *meeting point* between things and the forms of the human mind; phenomena are what *circulates* all along the reversible chain of transformations, at each step losing some properties to gain others that render them compatible with already-

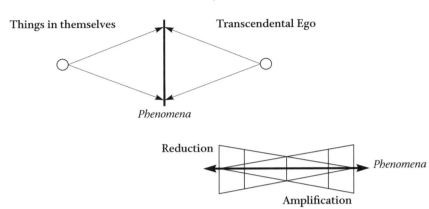

Figure 2.23 In the Kantian scenography, phenomena reside at the meeting point between the inaccessible things in themselves and the categorizing work made by the active Ego; with circulating reference, phenomena are what routinely circulates through the cascade of transformations.

established centers of calculation. Instead of growing from two fixed extremities toward a stable meeting point in the middle, the unstable reference *grows from the middle toward the ends,* which are continually pushed further away. To understand how Kantian philosophy has muddled the triangles, a fifteen-day expedition is all that is required. (All that is required, I hasten to add, on condition that I am not asked to speak of *my* work in the same lavish detail in which the pedologists report theirs: fifteen days would then become twenty-five years of hard labor at controversies with scores of dear colleagues equipped with decades worth of data, instruments, and concepts. I portray myself here, without fear of contradiction, as a simple spectator with easy access to the knowledge of my informants. A reflexivity that could follow every thread at once is, I would be the first to admit, beyond me.)

Is it possible, with the help of my schema, to understand, visualize, and detect why the original model of philosophers of language is so widespread, when this slightest inquiry quickly reveals its impossibility? Nothing could be simpler; all we need to do is obliterate, bit by bit, each of the stages we have witnessed in this photomontage (Figure 2.24).

Let us block in the extremities of the chain as if one were the refer-

Circulating reference

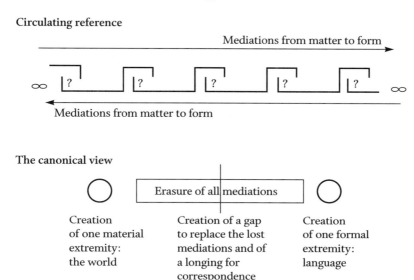

The canonical view

Figure 2.24 To obtain the canonical model of words and world separated by an abyss and related by the perilous bridge of correspondence, one has simply to consider the circulating reference and to eliminate all mediations as being unnecessary intermediaries that render the connection opaque. This is possible only at the (provisional) end of the process.

ent, the forest of Boa Vista, and the other were a phrase, "the forest of Boa Vista." Let us erase all the mediations that I have delighted in describing. In place of the forgotten mediations, let us create a radical gap, one capable of covering the huge abyss that separates the statement I utter in Paris and its referent six thousand kilometers away. *Et voilà,* we have returned to the former model, searching for something to fill the void we have created, looking for some *adequatio,* some resemblance between two ontological varieties that we have made as dissimilar as possible. It is hardly surprising that philosophers have been unable to reach an understanding on the question of realism and relativism: they have taken the two provisional extremities for the entire chain, as if they had tried to understand how a lamp and a switch could "correspond" to each other after cutting the wire and making the lamp "gaze out" at the "external" switch. As William James said in his powerful style:

The intermediaries which in their concrete particularity form a bridge, evaporate ideally into an empty interval to cross, and then, the relation of the end-terms having become saltatory, the whole hocus-pocus of *erkenntnistheorie* begins, and goes on unrestrained by further concrete considerations. The idea, in 'meaning' an object separated by an 'epistemological chasm' from itself, now executes what Professor Ladd calls a 'salto mortale' . . . The relation between idea and object, thus made abstract and saltatory, is thenceforward opposed, as being more essential and previous, to its own ambulatory self, and the more concrete description is branded as either false or insufficient. (James [1907] 1975, 247–248)

The next morning, after drafting the expedition report, we load the precious cardboard boxes containing the earthworms preserved in formaldehyde, and the neatly tagged little bags of earth, into the jeep (Figure 2.25). And this is what philosophical arguments that wish to link language to the world by a single regular transformation cannot successfully explain. From text we return to things, displaced *a little further*. From the restaurant-laboratory we set out for another laboratory a thousand kilometers away, in Manaus, and from there to Jussieu University in Paris, another six thousand kilometers away. Sandoval will return to Manaus alone with the precious samples that he must preserve intact despite the arduous trek that lies ahead. As I have said, each stage is matter for what follows and form for what precedes it, each separated from the other by a gap as wide as the distance between that which counts as words and that which counts as things.

They are getting ready to leave, *but they are also preparing to return.* Each sequence flows "upstream" and "downstream," and in this way the double direction of the movement of reference is amplified. To know is not simply to explore, but rather is to be able to make your way back over your own footsteps, following the path you have just marked out. The report that we drafted the night before makes this much clear: another expedition is required, to study the activity of those suspicious earthworms at the same field site:

From a pedological point of view, admitting that the forest is advancing on the savanna implies;
   1. that the forest and the biological activity particular to it transform a sandy soil into a clayey-sandy soil in the top 15 to 20 centimeters;

Figure 2.25

2. that this transformation would begin in the savanna in a 15- to 30-meter band at the edge.

While these two notions are difficult to conceive when starting from the assumptions of classical pedology, it is necessary, taking into account the solidity of the arguments derived from biological study, to test these hypotheses.

The clay enrichment of superior horizons cannot be accomplished by neoformation (lacking a known source of aluminum [aluminum is responsible for the creation of clay out of the silica contained in quartz]). The only agents capable of accomplishing this are the earthworms, whose activity on the studied site we have been able to verify, and which dispose of large quantities of koalinite contained in the horizon to a depth of 70 cm. The study of this worm population and the measure of its activity will therefore supply essential data for the continuation of this research.

Unfortunately, I will not be able to follow the next expedition. While the other members of the team say *au revoir* to Edileusa, I must say *adieu*. We are leaving by plane. Edileusa is staying in Boa Vista, pleased by an intense and friendly collaboration that was new to her, and she will continue to watch over her field site, which, because of the superposition of pedology and botany, has just increased in importance. And her plot will thicken more once we add the science of earthworms. Constructing a phenomenon in successive layers renders it more and more real within a network traced by the displacements (in both senses) of researchers, samples, graphics, specimens, maps, reports, and funding requests.

For this network to begin to lie—for it to cease to refer—it is sufficient to *interrupt* its expansion at either end, to stop providing for it, to suspend its funding, or to break it at any other point. If Sandoval's jeep swerves, breaking the jars of earthworms and scattering the little packages of earth, the whole expedition will have to be repeated. If my friends cannot find the funding to *return* to the field, we will never know if the sentence in the report about the role of the earthworms is a scientific truth, a gratuitous hypothesis, or a fiction. And if I lose all my negatives at the photo shop, how will anyone know whether I have lied?

Air conditioning at last! Finally, a space that looks more like a laboratory (Figure 2.26). We are in Manaus, at INPA, in an old work-

Figure 2.26

room transformed into an office. On the wall, Radambrasil's map of Amazonia and Mendeleev's chart. Offprints, files, slides, canteens, bags, cans of gasoline, an outboard motor. Smoking a cigarette, Armand writes the final version of the report on his laptop computer.

The forest-savanna transition of Boa Vista continues its transformations. Once typed in and saved on disk, the transition will circulate by fax, electronic mail, diskette, preceding the suitcases heavy with the earth and earthworms that will undergo various series of new trials in various laboratories selected by our pedologists. The results will return to thicken the piles of notes and files on Armand's desk, in support of his request for funding to return to the field. The unending round of scientific credibility: each turn absorbs more of Amazonia into pedology, a motion that cannot stop lest significance and signification be immediately lost.

Smoking a cigar, I too am writing my report on my laptop. Back in Paris, I am sitting at a desk cluttered with books, files, and slides, in front of an immense map of the Amazon basin. Like my colleagues, I extend the network of the forest-savanna transition—all the way to philosophers and sociologists, to the readers of this book. The section of the network that I am constructing, however, is made, *not* of the sort of references enacted by the other scientists, but of allusions and illustrations. My schemas do not refer in the same way as their diagrams and maps. Unlike Armand's inscription of the soil of Boa Vista, my photographs do not transport that of which I speak. I am writing a text of empirical philosophy that does not re-represent its evidence in the manner of my pedologist friends, and hence the traceability of my subject matter is not sufficiently immutable to permit the reader's return to the field. (I will leave it to the reader to measure the distance that separates the natural and social sciences, for that mystery would require another expedition, one that would study the role of the bantam empiricist that I have been playing.)

You can now look at a map of Brazil in an atlas, at the area around Boa Vista, but not for a *resemblance* between the map and the site whose story I have been recounting. This whole tired question of the correspondence between words and the world stems from a simple confusion between epistemology and the history of art. We have taken science for realist painting, imagining that it made an exact copy of the world. The sciences do something else entirely—paintings too, for that matter. Through successive stages they link us to an aligned, transformed, constructed world. We forfeit resemblance, in this model, but there is compensation: by pointing with our index fingers to features of an entry printed in an atlas, we can, through a series of uniformly discontinuous transformations, link ourselves to Boa Vista. Let us rejoice in this long chain of transformations, this potentially endless sequence of mediators, instead of begging for the poor pleasures of *adequatio* and for the rather dangerous *salto mortale* that James so nicely ridiculed. I can never verify the resemblance between my mind and the world, but I can, if I pay the price, *extend* the chain of transformations wherever verified reference circulates through constant substitutions. Is this "deambulatory" philosophy of science not more realist, and certainly more *realistic,* than the old settlement?

Science's Blood Flow

An Example from Joliot's Scientific Intelligence

Now that we have begun to understand that reference is something that circulates, everything is going to change in our understanding of the connections between a scientific discipline and the rest of its world. In particular, we are going to be able to reconnect many of the contextual elements that we had to abandon in the previous chapter. With more than a little exaggeration, science studies can be said to have made a discovery not totally dissimilar to that of the great William Harvey himself . . . By following the ways in which facts circulate, we will be able to reconstruct, blood vessel after blood vessel, the whole circulatory system of science. The notion of a science isolated from the rest of the society will become as meaningless as the idea of a system of arteries disconnected from the system of veins. Even the notion of a conceptual "heart" of science will take on a completely different meaning once we begin to examine the rich vascularization that makes the scientific disciplines alive.

To exemplify this second point I will take a canonical example, this time not from a science as green and friendly as pedology but from one as heavy and somber as atomic physics. My intention is not to add to the history and anthropology of physics as many of my colleagues have so excellently done (Schaffer 1994; Pickering 1995; Galison 1997), but to recast the meaning of the little adjective "social." If, in Chapter 2, I had to abandon most of the threads leading outward to the context of the expedition, in this chapter I will leave out most of the technical content to concentrate on the *threading* itself. This will allow me to introduce the little bit of classic sociology of science that we need to continue, and to help readers who believe science studies aims to provide a "social" explanation of science abandon this prejudice. Once we are equipped with a different notion of reference and a renewed conception of the social, it will be possible to integrate the two with an alternative definition of the object. I wish I could go faster, but going fast, in these matters, is a sure recipe for simply repeating the old settlement without any hope of illuminating the new one that is still cloaked in darkness.

A Little Example from Joliot

In May 1939 Frédéric Joliot, advised by his friends in the Ministry of War and by André Laugier, the director of the recently established CNRS (Centre National de la Recherche Scientifique, France's National Center for Scientific Research), entered into a very subtle legal agreement with a Belgian company, the Union Minière du Haut-Katanga. Thanks to the discovery of radium by Pierre and Marie Curie and the discovery of uranium deposits in the Congo, this company had become the most important supplier to all the laboratories in the world that were feeling their way toward the production of the first artificial nuclear chain reaction. Joliot, like his mother-in-law Marie Curie before him, had found a way of getting the company involved. In fact, the Union Minière used its radioactive ores only as a source of radium, which it sold to doctors; immense heaps of uranium oxide were left lying about at its waste sites. Joliet planned to build an atomic reactor, for which he would need a huge quantity of uranium; this made what had been a mere waste product of the production of radium into something valuable. The company promised Joliot five tons of uranium oxide, technical assistance, and a million francs. In return, all the French scientists' discoveries would be patented by a syndicate which would distribute the profits fifty-fifty between the Union Minière and the CNRS.

Meanwhile, in his laboratory at the Collège de France, Joliot and his two main research colleagues, Hans Halban and Lew Kowarski, were looking for an arrangement just as subtle as the one that had brought together the interests of the Ministry of War, the CNRS, and the Union Minière. But this time it was a matter of coordinating the apparently irreconcilable behaviors of atomic particles. The principle of fission had just been discovered. When bombarded by neutrons, each atom of uranium broke in two, liberating energy. This artificial radioactivity had a consequence that was immediately grasped by several physicists: if under bombardment each atom of uranium gave off two or three other neutrons which in turn bombarded other atoms of uranium, an extremely powerful chain reaction would be set in motion. Joliot's team immediately set to work to prove that such a reaction could be produced, and that it would open the way to new scientific discoveries and to a new technique for producing energy in unlimited quantities. The first team able to prove that each generation of neutrons did indeed give birth to an even greater number would gain considerable prestige in the highly competitive scientific community, in which the French occupied, at that time, a position of the first rank.

Determined to pursue this important scientific discovery, Joliot and his colleagues continued to publish their findings, despite the urgent telegrams Leo Szilard was sending them from America. In 1934 Szilard, an émigré from Hungary and a visionary physicist, had taken out a secret patent on the principles of construction of an atomic bomb. Worried that the Germans too would develop an atomic bomb as soon as they could be certain that the neutrons emitted were more numerous than those present at the beginning, Szilard fought to encourage self-censorship by all anti-Nazi researchers. He could not, however, prevent Joliot from publishing a final article in the English journal *Nature* in April 1939, which showed that it might be possible to generate 3.5 neutrons per fission. On reading this article, physicists in Germany, England, and the Soviet Union all had the same thought: they immediately reoriented their research toward bringing about a chain reaction and just as quickly wrote to their governments to alert them to the vast importance of this research, to inform them of its dangers, and to request immediate provision of the enormous resources needed to test Joliot's claim.

Around the world about ten different teams became passionately engaged in the attempt to produce the first artificial nuclear chain reaction, but only Joliot and his team were already in a position to turn this into an industrial or military reality. Joliot's first problem was to slow down the neutrons emitted by the first fissions, for if these were too fast they would not set off the reaction. The team looked for a moderator that could slow the neutrons without absorbing them or bouncing them back; thus the ideal moderator would have a set of properties very difficult to reconcile. In their workshop at Ivry, they tried different moderators under different configurations, for example paraffin and graphite. It was Halban who drew their attention to the decisive advantages of deuterium, an isotope of hydrogen, twice as heavy but with the same chemical behavior. It could take the place of hydrogen in water molecules, which then became "heavy." From earlier work he had done on heavy water, Halban knew that it absorbed very few neutrons. Unfortunately, this ideal moderator had one major drawback: there was only one atom of deuterium in every 6,000 atoms of hydrogen. It cost a fortune to obtain heavy water, and it was produced on an industrial scale at only one plant in the world, which belonged to the Norwegian company Norsk Hydro Elektrisk.

Raoul Dautry, a graduate of the Ecole Polytechnique and a senior civil servant who became the French Minister of Armaments only too shortly before the defeat of France in World War II, was also kept informed of Joliot's work from the very beginning. He had been in favor of Joliot's agreement with the Union Minière and did everything he could to support the team at the Collège de France and the early days of the CNRS, attempting to integrate, as much as French tradition allowed, military and advanced scientific research. Although he did not share Joliot's leftist political opinions, he had the same confidence in the progress of knowledge and the same passion for national independence. Joliot promised an experimental reactor for civilian use which might eventually lead to the construction of a new type of armament. Dautry and other technocrats offered Joliot generous support while asking him to change his priorities: if the bomb was practicable, it must be developed first and very quickly.

Halban's calculations on the slowing of neutrons, Joliot's hypothesis of the feasibility of the chain reaction, and Dautry's conviction about the necessity of developing new armaments became even more closely entwined when it came to obtaining the heavy water from Norway. While the "phony war" was taking place between the Siegfried and the Maginot lines, spies, bankers, diplomats, and German, English, French, and Norwegian physicists fought over twenty-six containers the Norwegians had given the French to prevent the Germans from getting hold of them. After an eventful few weeks the containers reached Joliot's possession. Halban and Kowarski, both foreigners and therefore suspect, had been put out to pasture by the French secret service for the duration of the operation. Once it was completed, they were authorized to return to the laboratory at the Collège de France, where under the protection of Dautry and the military, they set to work to combine the uranium from the Union Minière and the heavy water from the Norwegians with the calculations that Halban worked out every day with the confusing data from their primitive Geiger counter.

How to Link the History of Science with That of France

How should we understand this story, so well told by the American historian Spencer Weart (1979), of which I have given only a summary of a single episode? Two major misunderstandings have made the project of science studies of mapping the circulatory system of science incomprehensible. The first is the belief that science studies seeks a "social explanation" of scientific facts; the second the belief that it deals only with discourse and rhetoric, or at best epistemological questions, but does not care about "the real world outside." Let us clear up each of these misunderstandings in turn.

Science studies, to be sure, rejects the idea of a science disconnected from the rest of society, but this rejection does not mean that it embraces the opposite position, that of a "social construction" of reality, or that it ends up in some intermediary position, trying to sort out "purely" scientific factors from "merely" social ones (see the end of Chapter 4). What science studies rejects is *the entire research program* that would try to divide the story of Joliot into two parts: one for the legal problems with the Union Minière, the "phony war," Dautry's nationalism, the German spies; and the other for neutrons, deuterium, the absorption coefficient of paraffin. A scholar of this period would then have two lists of characters corresponding to two stories: in the first, the history of France from 1939 to 1940; in the second, the history of science in the same period. The first list would deal with politics, law, economics, institutions, and passions; the other with ideas, principles, knowledge, and procedures.

We might even imagine two subprofessions, two different kinds of historians, one preferring explanations by pure politics, the other by pure science. The first kind of explanation is usually called *externalist** and the second *internalist**. In this period of 1939–1940, these two histories would have no points of intersection. The one would speak of Adolf Hitler, Raoul Dautry, Edouard Daladier, and the CNRS, but not of neutrons, deuterium, or paraffin; the other would talk about the principle of the chain reaction, but not about the Union Minière or the banks that owned Norsk Hydro Elektrisk. Like two teams of civil engineers working in two parallel valleys in the Alps, they both would do an enormous amount of work without ever knowing of each other's existence.

Of course, once this division between human and nonhuman actors was drawn, everyone would admit that there remained a slightly muddled area of hybrids, which might be found perhaps in one column, perhaps in the other, or perhaps in neither. To deal with this "twilight zone," externalists and internalists would have to borrow factors from each other's lists. One might say, for example, that Joliot "mixed up" political concerns with purely scientific interests. Or one might say that the plan to slow neutrons with deuterium was, of course, a scientific project, but that it was also "influenced" by extrascientific factors. Szilard's project of self-censorship was not "strictly scientific," we might say, because it introduced military and political considerations into the free interchange of ideas of pure science. In this way, everything that appears mixed is explained by reference to one of two equally pure constituents: politics and science.

Science studies could be defined as the project whose aim is to do away with this division altogether. The story of Joliot as told by Spencer Weart is a "seamless web" which cannot be torn in two without making both the politics of the time and the atomic physics incomprehensible. Instead of following the parallel valleys, the purpose of science studies is to dig a tunnel between them by putting together two teams, which attack the problem from opposite ends and hope to meet in the middle.

By following Halban's arguments on cross-sections (Weart 1979), which conclude that deuterium has decisive advantages, the analyst of science is led, without prejudice and without postulating a great divide between science and politics, through an imperceptible *transition* into Dautry's office, and from there into the plane of Jacques Allier, a banker and flying officer who was the secret agent sent by France to outwit the fighters of the Luftwaffe. Starting on the science side of the tunnel, the historian ultimately arrives on the other side, with war and politics. But en route she might meet a colleague coming from the other direction who started with the industrial strategy of the Union Minière and, through another imperceptible transition, ended up very interested in the method of extraction of uranium 235, and subsequently in Halban's calculations. Starting from the politics side, this historian, willingly or not, becomes involved in mathematics. Instead of two histories which do not intersect at any point, we now have people who tell two symmetrical stories which include the *same* elements and the *same* actors, but *in the opposite order*. The first scholar expected to follow Halban's calculations without having to deal with the Luftwaffe, and the second imagined that he could look at the Union Minière without having to do any atomic physics.

They were both mistaken, but the paths they traced, thanks to the opening of the tunnel, are much more interesting than they had expected. In fact, by following without prejudice the interconnected threads of their reasoning, science studies will reveal *a posteriori* the work the scientists and the politicians had to do to become so inextricably bound together. It wasn't determined in advance that all the elements of Weart's account should be mixed together. The Union Minière could have carried on producing and selling copper without bothering about radium or uranium. If Marie Curie and later Frédéric Joliot had not worked at getting the company interested in the work done in their laboratories, an analyst from the Union Minière would never have had to do nuclear physics. When discussing Joliot, Weart would never have had to speak of the Upper Katanga. Conversely, once he had envisioned the possibility of a chain reaction, Joliot could have directed his research at some other topic, without having to mobilize, in order to produce a reactor, nearly all of France's industrialists and enlightened technocrats. Writing about prewar France, Weart would not have had to mention Joliot.

In other words, the project of science studies, contrary to what science warriors have tried to make everyone believe, is not to state *a priori* that there exists "some connection" between science and society, because *the existence of this connection depends on what the actors have done or not done to establish it.* Science studies merely provides the means of tracing this connection *when it exists*. Instead of cutting the Gordian knot—on the one hand pure science, on the other pure politics—it struggles to follow the gestures of those who tie it tighter. The social history of the sciences does not say: "Look for society hidden in, behind, or underneath the sciences." It merely asks some simple questions: "In a given period, how long can you follow a policy before having to deal with the detailed content of a science? How long can you examine the reasoning of a scientist before having to get involved with the details of a policy? A minute? A century? An eternity? A second? All we ask of you is not to cut away the thread when it leads you, through a series of imperceptible transitions, from one type of element to another." All the answers are interesting and count as major data for anyone who wishes to understand this imbroglio of things and people—*including*, of course, the data that might show that there is not the slightest connection, at a given time, between a piece of science and the rest of the culture.

It is not enough to say that the connections between science and politics form a very tangled web. To refuse any *a priori* division between the list of human or political actors and that of ideas and procedures is no more than a first step, and an entirely negative one at that. We must also be able to understand the series of operations by which an industrialist who wanted only to develop his business found himself forced to do calculations of the rate of absorption of neutrons by paraffin; or how someone who wanted nothing but a Nobel Prize set about organizing a commando operation in Norway. In both cases the *initial* vocabulary is different from the *final* vocabulary. There is a *translation** of political terms into scientific terms and vice versa. For the managing director of the Union Minière, "making money" now means, to some extent, "investing in Joliot's physics"; while for Joliot, "demonstrating the possibility of a chain reaction" now means in part "looking out for Nazi spies." The analysis of these translation operations makes up a large portion of science studies. The idea of translation provides the two teams of scholars, one coming from the side of politics and going toward the sciences and the other coming from the side of the sciences and following the circulating references, with the system of guidance and alignment that gives them some chance of meeting in the middle rather than missing each other.

Let us follow an elementary operation of translation, so as to understand how in practice one passes from one register to another. Dautry wants to ensure France's military strength and the self-sufficiency of its energy production. Let us say that this is his "goal"—whatever psychology we wish to impute to him. Joliot wants to be the first in the world to produce controlled artificial nuclear fission in the laboratory; this is his goal. To call the first ambition "purely political" and the second "purely scientific" is completely pointless, because it is the "impurity" alone that will allow both goals to be attained.

Indeed, when Joliot met Dautry he did not particularly try to change Dautry's goal, but to position his own project in such a way that Dautry would see the nuclear chain reaction as the *fastest* and most certain way of achieving national independence. "If you use my laboratory," Joliot may have said, "it will be possible to gain a significant lead over other countries, and perhaps even to produce an explosive that goes beyond anything we know." This transaction is not of a commercial nature. For Joliot it is not a question of selling nuclear fission, since it doesn't even exist yet. On the contrary, the only way he can make it exist is to receive from the Minister of Armaments the personnel, the premises, and the connections that will enable him, in the middle of a war, to obtain the tons of graphite, the uranium, and the liters of heavy water that are needed. Both men believe that, since it is impossible for either to achieve his goal directly, political and scientific purity are in vain, and that it will thus be best to negotiate an arrangement that modifies the relation between their two original goals.

The operation of translation consists of combining two hitherto different interests (waging war, slowing down neutrons) to form a single composite goal (see Figure 3.1). Of course there is no guarantee that one or the other of the parties isn't cheating. Dautry may be squandering precious resources by letting Joliot fool around with his neutrons while the Germans are massing their tanks in the Ardennes. Conversely, Joliot may feel he is being forced to build the bomb before the civilian reactor. Even if the balance is equal, neither of the parties, as is shown in the diagram, will be able to arrive at *exactly* his original goal. There is a drift, a slippage, a displacement, which, depending on the case, may be tiny or infinitely large.

In the case we are using as an example, Joliot and Dautry did not achieve their goals until fifteen years later, after a terrible defeat, when General de Gaulle created the CEA, the Commissariat à l'Énergie

Before translation

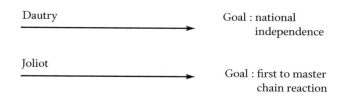

After translation

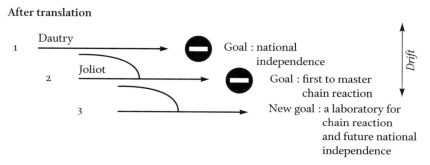

Figure 3.1 One should be careful not to fix interests *a priori;* interests are "translated." That is, when their goals are frustrated, actors take detours through the goals of others, resulting in a general drift, the language of one actor being substituted for the language of another.

Atomique (Atomic Energy Commission). What is important in such an operation of translation is not only the fusion of interests that it allows but the creation of a new mixture, the laboratory. In fact, the shed at Ivry became the crucial juncture that would allow the joint realization of both Joliot's scientific project and the national independence so close to Dautry's heart. The laboratory's walls, its equipment, its staff, and its resources were brought into existence by both Dautry and Joliot. It was no longer possible to tell, among the complex of forces mobilized around the copper sphere filled with uranium and paraffin, what belonged to Joliot and what to Dautry.

To study a single negotiation or translation in isolation would be useless. Joliot's labors could not of course be confined to ministerial offices. Having gained his laboratory, he now had to go and negotiate *with the neutrons themselves.* Was it one thing to persuade a minister to provide a stock of graphite, and quite another to persuade a neutron to slow down enough to hit a uranium atom so as to provide three more neutrons? Yes and no. For Joliot it wasn't very different. In the morning he dealt with the neutrons and in the afternoon he dealt with the minister. The more time passed, the more these two problems became one: if too many neutrons escaped from the copper vessel and lowered the output of the reaction, the minister might lose patience. For Joliot, containing the minister and the neutrons in the same project, keeping them acting and keeping them under discipline, were not really distinct tasks. *He needed them both.*

Joliot crossed and recrossed Paris, moving from mathematics to law and to politics, sending telegrams to Szilard so the flow of publications needed to promote the project would continue, telephoning his legal adviser so the Union Minière would keep sending uranium, and recalculating for the *n*th time the absorption curve obtained with his rudimentary Geiger counter. Such was his scientific work: holding together all the threads and getting favors from everybody, neutrons, Norwegians, deuterium, colleagues, anti-Nazis, Americans, paraffin . . . No one said being a scientist was a simple job! To be *intelligent*, as the word's etymology indicates, is to be able to hold all these connections at once. To understand science is, with Joliot's help (and Weart's), to understand this complex web of connections without imagining in advance that there exist a given state of society and a given state of science.

It is now easier to see the difference between science studies and the two *parallel* histories that it replaces. In order to explain all the political and scientific imbroglios, the two teams of historians always had to see them as regrettable intermixings of two equally pure registers. All their explanations therefore had to be couched in terms of "distortion," of "impurity," or at best of "juxtaposition." For them, purely political or economic factors were added to purely scientific ones. Where these historians saw only confusion, science studies sees a slow, continuous, and entirely explicable *substitution* of a certain kind of concern and a certain kind of practice for another. There are in fact moments when, if one holds firmly the calculation of the cross-section of deuterium, *one also holds*, through substitutions and translations, the fate of France, the future of industry, the destiny of physics, a patent, a good paper, a Nobel Prize, and so on.

With the help of another diagram it is possible to extend the contrast between these two types of inquiry into the connections of science. The left side of Figure 3.2 portrays the separation between science and politics in its most common form: there is a nucleus of scientific content *surrounded* by a social, political, and cultural "environment," which can be called the "context" of science. On the basis of such a separation it is possible to offer either externalist explanations or internalist ones and to feed the contradictory research programs of our two teams of scholars. The members of the first team will use the vocabulary of context* and will attempt (sometimes) to penetrate as far as they can into the scientific content; the members of the second will use the vocabulary of content* and will remain within the central conceptual core. For the first, *what explains science is society*—although usually only the surface of the discipline is in question: its organization, the relative status of different workers, or the errors it is later shown to have produced. In the second the *sciences explain themselves,* without need of external assistance since they provide their own commentary about themselves and develop from their own inner forces. To be sure, the social environment can either hinder or encourage their development, but it never forms or constitutes the very content of the sciences.

On the right side of Figure 3.2 is the science studies program that can be called the translation model* (Callon 1981). It should now be clear that there is no relation between the two paradigms. Science studies does *not* occupy a position inside the classical debate between internalist and externalist history. It entirely reconfigures the questions. The only thing one can say is that the successive chains of translation involve, at one end, *exoteric* resources (which are more like what we read about in the daily papers), and at the other end, *esoteric* resources (which are more like what we read in university textbooks). But these two ends are no more important and no more real than the two extremities of reference in the previous chapter, and for the same reason. Everything important happens *between the two,* and the same explanations serve to carry the translation in both directions. In this second model the same methods are used to understand science and society. Science studies has never had any interest whatsoever, at least as I see it, in providing a social explanation of any given piece of science. If it had, it would have failed immediately, since *nothing* in the

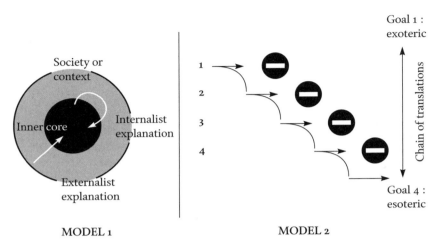

Figure 3.2 In Model 1 science is conceived as a core surrounded by a corona of social contexts that are irrelevant to the definition of science; thus internalist and externalist explanations have little in common. In Model 2 the successive translations have made esoteric and exoteric vocabularies have something in common, and the distinction between internalist and externalist explanations is exactly as small (or as large) as the chain of translation itself.

ordinary definition of what society is could account for the connection between a Minister of Armaments and neutrons. It is only because of Joliot's work that this connection has been made. Science studies follows those implausible translations which mobilize in completely unexpected ways fresh definitions of what it is to make war and fresh definitions of what the world is made of.

The Progressive Packing of Nonhumans into Human Discourse

Now that the first misunderstanding has been cleared up, it will be easier to deal with the second, especially with the help of what we learned about circulating reference in Chapter 2. Scientists not only blur, in their daily practice, the boundary between their pure esoteric science and the impure exoteric realm of society, they also blur the boundary between the domain of discourse and what the world is like.

Philosophers of science like to remind us, as if this were the epitome of good common sense, that we should never confuse epistemological questions (what our representation of the world is) and ontological questions (what the world is really like). Unfortunately, if we followed the philosophers' advice we would not understand any scientific activity, since confusing those two supposedly separate domains is precisely what scientists spend much of their time doing. Joliot not only translates social and scientific considerations more and more intimately, he also mixes up epistemological and ontological questions more thoroughly every day. It is only because of this gradually accumulating confusion that what he says about chain reactions can be taken more and more seriously by others.

Consider this sentence: (1) "Every neutron liberates 2.5 neutrons." This is what one reads today in encyclopedias. This is what is called a "scientific fact." Now let us take another sentence: (2) "Joliot claims that each neutron liberates 3 to 4 neutrons, but that's impossible; he has no proof; he's far too optimistic; that's the French all over, counting their chickens before they're hatched; and in any case, it's incredibly dangerous; if the Germans read his claim, they'll believe it's all possible and work on it seriously." Unlike sentence (1), sentence (2) does not conform to the stylistic rules governing the appearance of scientific facts: it cannot be read in any encyclopedia. Its dated character is easily discernible (somewhere between 1939 and 1940) and it might be ascribed to a fellow physicist (such as Szilard, who had found a haven at that time in Enrico Fermi's laboratory on the South Side of Chicago). We may note that these two sentences have a section in common, the statement or *dictum*:* "each neutron liberates *x* neutrons"; and a very different part, made up of an ensemble of situations, people, and judgments, called the modifier or *modus*.*

As I have shown many times, a convenient marker of the appearance of a scientific fact is that the modifier drops entirely and only the dictum is maintained. The elimination of these modifiers is the result and sometimes the goal of scientific controversy (as we will see in Chapter 4, when Pasteur steps back from his yeast cells to let them speak for themselves). For example, if Joliot and his group have done their work successfully, his colleagues will move imperceptibly from the second sentence to a third, more respectful one: (3) "The Joliot team seems to have proved that every neutron liberates three neutrons; that's very interesting." A few years later we will read sentences like this: (4) "Numerous experiments have proven that each neutron liberates between 2 and 3 neutrons." One more effort, and we arrive at the phrase with which we started: (1) "Each neutron liberates 2.5 neutrons." A little later this sentence, without a trace of qualification, without author, without judgment, without polemics or controversies, without even any allusion to the experimental mechanism that made it possible, will enter into a state of even greater certainty. Atomic physicists will not even speak of it, will even stop writing it—except in an introductory course or a popular article—so obvious will it have become. From lively controversy to tacit knowledge, the transition is progressive and continuous—at least when everything goes well, which is, of course, very rarely.

How are we going to account for this progressive shift from (2) to (1) through (3) and (4)? Are we going to say, to use the tired cliché, that they tend "asymptotically" toward the true state of affairs? Are we going to say that (2) is still a human statement marked by language and history while (1) is not a statement at all and has escaped history and humanity altogether? The traditional way to answer these questions is to try to identify among these statements the ones that correspond to a state of affairs and the ones that have no reference. But again, science studies is not the research program that would take a position in this classical debate. As we saw in Chapter 2, it is interested in a rather different question: how can the world be progressively packed into discourse through successive transformations so that a stable flow of reference in two directions may ensue? How can Joliot get rid of the qualifications that hedge the scientific fact he wishes to establish? The answer to this question explains why there can be no other history of science than science studies, as I am defining it here.

Joliot may be convinced in his own mind that the nuclear chain reaction is feasible and that it will lead in a few years to the construction of an atomic reactor. If, however, each time he states this possibility, his colleagues add qualifications—such as "It is ridiculous to believe that [dictum]," "It is impossible to think that [dictum]," "It is dangerous to imagine that [dictum]," "It is contrary to theory to claim that [dictum]"—Joliot will find himself utterly powerless. He cannot *by himself* transform the statement he is proposing into a scientific fact that the others accept; by definition, he needs *the others* to bring about this transformation. It was Szilard who had to admit, "I am now convinced that Joliot can make his reactor work," even if he immediately added, "as long as the Germans don't get ahold of it if they occupy Paris." In other words, to reuse a slogan I have often employed, the statement's fate is in the hands of others, in the hands mostly of dear colleagues, who are for this reason both loved and hated (the fewer they are, and the more esoteric or important the statement in question, the more they will be loved or hated).

I am not trying to stress here the regrettable "social dimension" of science that would serve to prove that scientists are only human, all too human. Controversy is not something that would disappear if researchers would only be "really scientific." There is no way to skip any of the steps toward conviction; one might as well imagine Joliot immediately writing an encyclopedia article on the operation of a nuclear power plant! It is always necessary to convince the others first, one by one. The others are always there, skeptical, undisciplined, inattentive, uninterested; they form the social group that Joliot cannot do without.

Joliot, like all researchers, needs the others, needs to discipline them and to convince them; he is not able to do without them and lock himself up in the Collège de France, alone with his firm conviction that he is right. He is not, however, completely without weapons of his own. Despite the slanderous claim of the science warriors, science studies has never said that the "others" mixed up in the conviction process were all humans. On the contrary, the whole effort of science studies has been to follow the extraordinary mixtures of humans and nonhumans that scientists had to devise in order to convince. Into his discussions with colleagues Joliot can introduce *other resources* than the ones classically handed down to him by rhetoric.

This is the very reason he was in such a hurry to slow down the neutrons with deuterium. Alone, he could not force his colleagues to believe him. If he could get his reactor going for only a few seconds, and if he could get evidence of this event that was sufficiently clear that no one could accuse him of seeing only what he wanted to see, then Joliot would no longer be alone. With him, behind him, disciplined and supervised by his collaborators, and properly lined up, the neutrons of the reactor could be made visible in the form of a cross-sectional diagram. The experiment in the shed at Ivry was very expensive, but it was precisely this expense that would force his esteemed colleagues to take his article in *Nature* seriously. Science studies, once again, does not take a position in a classical debate—is it rhetoric or proof that finally convinces scientists?—but reconfigures the whole question in order to understand this strange hybrid: a copper sphere built to convince.

For six months Joliot was the only one in the world who had at his disposal the material resources allowing him to mobilize both colleagues and neutrons around and inside a real reactor. Joliot's opinion by itself could be swept aside with a wave of the hand; Joliot's opinion supported by Halban's and Kowarski's diagrams, diagrams obtained from the copper sphere in the shed at Ivry, could not so easily be cast aside—the proof being that three countries at war immediately set to work at building their own reactors. Disciplining men and mobilizing things, mobilizing things by disciplining men; this is a new way of convincing, sometimes called scientific research.

In no way is science studies an analysis of the rhetoric of science, of the discursive dimension of science. It has always been an analysis of how language slowly becomes capable of transporting things themselves *without* deformation *through* transformations. The notion of the huge gap between words and world made it impossible to understand this progressive loading—as did the very distinction between rhetoric and reality, the political origins of which I will examine in Chapter 7. But getting rid of a nonexistent gap and of an even less real correspondence between two nonexistent things—words and world—is not at all the same thing as saying that humans are forever stuck in the prison of language. It implies exactly the opposite. Nonhumans can be loaded into discourse exactly as easily as ministers can be made to understand neutrons. As we will see in Chapter 6, this is the simplest of all things to do. Only the sway of the modernist settlement could make this commonsensical evidence appear bizarre.

What seemed shocking at first in this new paradigm was that it did not rely on the myth of a heroic break *away* from society, convention, and discourse, a mythical break that would let the solitary scientist discover the world as it is. To be sure, we no longer portray scientists as those who abandon the realm of signs, politics, passions, and feelings in order to discover the world of cold and inhuman things in themselves, "out there." But that does not mean that we portray them as talking to humans, to humans only, because those they address in their research are not exactly humans but strange hybrids with long tails, trails, tentacles, filaments tying words to things which are, so to speak, *behind* them, accessible only through highly indirect and immensely complex mediations of different series of instruments. The truth of what scientists say no longer comes from their breaking away from society, convention, mediations, connections, but from the safety provided by the circulating references that cascade through a great number of transformations and translations, modifying and constraining the speech acts of many humans over which no one has any durable control. Instead of abandoning the base world of rhetoric, argumentation, calculation—much like the religious hermits of the past—scientists begin to speak in truth because they plunge even more deeply into the secular world of words, signs, passions, materials, and mediations, and extend themselves ever further in intimate connections with the nonhumans they have learned to bring to bear on their discussions.

If the traditional picture had the motto "The more disconnected a science the better," science studies says, "The more connected a science, the more accurate it may become." The quality of a science's reference does not come from some *salto mortale* out of discourse and society in order to access things, but depends rather on the extent of its transformations, the safety of its connections, the progressive accumulation of its mediations, the number of interlocutors it engages, its ability to make nonhumans accessible to words, its capacity to interest and to convince others, and its routine institutionalization of these flows (see Chapter 5). There do not exist true statements that correspond to a state of affairs and false statements that do not, but only continuous or interrupted reference. It is not a question of truthful scientists who have broken away from society and liars who are influenced by the vagaries of passion and politics, but one of highly connected scientists, such as Joliot, and sparsely connected scientists limited only to words.

The imbroglio with which this chapter began is not a regrettable aspect of scientific production; it is the result of that very production. At every point one finds people and things mixed up, opening up a controversy or putting an end to one. If, after Joliot had outlined his project, Dautry had not received a favorable response from his advisers,

Joliot would not have obtained the resources to mobilize the tons of graphite his experiment demanded—and if he had not been able to convince Dautry's advisers, he would not have been able to convince his own colleagues. It is the same scientific work that led him to go down to the shed at Ivry, to go up to Dautry's office, to approach his colleagues, to go back over his calculations. It is this same disciplining and disciplined labor that led him to concern himself with the development of the CNRS—without which he would not have had colleagues sophisticated enough in the new physics (Pestre 1984) to find his arguments interesting; to give lectures to the workers in the Communist suburbs—without which there would not have been widespread support for scientific research as a whole; to get the directors of the Union Minière to visit his laboratory—without which he would not have received the tons of radioactive waste needed for his reactor; to write articles for *Nature*—without which the very goal of his research would have been foiled; but, above all else, to struggle to get the damned reactor working.

As we will see, the energy with which Joliot pushed Szilard, Kowarski, Dautry, and all the others is *proportional* to the number of resources and interests he had already mobilized. If the reactor fails, if each neutron liberates no more than one other neutron, then all these accumulated resources will scatter and disperse. It will no longer be worth going to all this trouble. This line of research will be seen as costly, useless, or premature, and Joliot's words will begin to lie, to lack reference. What matters for science studies is that a heterogeneous assembly of hitherto unrelated elements now shares a common fate within a common collective, and that Joliot's words will become true or false according to what circulates throughout this entire newly assembled collective. It is too late to claim that ontological and epistemological questions should be kept neatly distinct. Because of Joliot's work these questions are now tied to one another, and the relevance of what he says to what the world is really like now hinges on what happens in the copper sphere at Ivry.

The Circulatory System of Scientific Facts

Translation operations transform political questions into questions of technique, and vice versa; during a controversy, operations of conviction mobilize a mixture of human and nonhuman agents. Instead of defining *a priori* the distance between the nucleus of scientific content and its context, an assumption that would render incomprehensible the numerous short-circuits between ministers and neutrons, science studies follows leads, nodes, and pathways no matter how crooked and unpredictable they may look to traditional philosophers of science. If it is impossible, by definition, to give a general description once and for all of the unpredictable and heterogeneous links that explain the circulatory system that keeps scientific facts alive, it must nevertheless be possible to outline the different preoccupations that all researchers will hold simultaneously if they want to be good scientists.

Let us try to enumerate the various flows that Joliot must take into account simultaneously and that together guarantee the reference for what he says. All at the same time, Joliot must get the reactor to work; convince his colleagues; interest the military, politicians, and industrialists; give the public a positive image of his activities; and, last but not least, understand what is going on with these neutrons that have become so important to the parties he has interested in their fate. These are *five* types of activities that science studies needs to describe first if it seeks to begin to understand in any sort of realistic way what a given scientific discipline is up to: instruments, colleagues, allies, public, and finally, what I will call *links* or *knots* so as to avoid the historical baggage that comes with the phrase "conceptual content." Each of these five activities is as important as the others, and each feeds back into itself and into the other four: without allies, no graphite, and thus no reactor; without colleagues, no favorable opinion from Dautry, and thus no expedition to Norway; without a way of calculating the neutrons' rate of reproduction, no assessment of the reactor, so no proof, and thus no colleagues convinced. In Figure 3.3 I have mapped the five different loops that science studies needs to consider in order to reconstruct the circulation of scientific facts.

Mobilization of the World

The first loop one has to follow can be called the *mobilization of the world*, if we understand by this very general expression all the means by which nonhumans are progressively loaded into discourse, as we

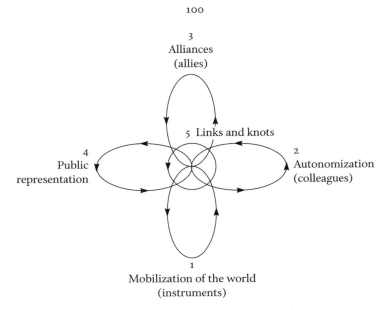

Figure 3.3 Once we abandon the core/context model, it is possible to deploy an alternative one. Five loops have to be taken into account simultaneously for any realistic rendering of science; in this model, the conceptual element (links and knots) is still in the middle, but it is situated more like a central knot tying the four other loops than like a stone surrounded by a context.

saw in Chapter 2. It is a matter of moving toward the world, making it mobile, bringing it to the site of controversy, keeping it engaged, and making it available for arguments. In certain disciplines, such as Joliot's nuclear physics, this expression primarily designates the *instruments* and major *equipment* that, at least since World War II, have made up the history of Big Science. For many other disciplines it will also designate the *expeditions* sent around the world over the past three or four centuries to bring back plants, animals, trophies, and cartographical observations. We saw an example of this in Chapter 2, with the soil of the Amazon forest becoming more and more mobile and beginning a long voyage, through a series of transformations, toward the University of Paris. For still other disciplines the word "mobilization" will mean neither instruments, equipment, nor expeditions, but *surveys*, the questionnaires that have gathered information about the state of a society or an economy.

Whatever the kinds of mediations put to work, this loop is concerned with doing practically what Kant called a Copernican Revolution, though he hardly realized how very practical was the activity designated by this grandiose expression: instead of moving around the objects, scientists make the objects move around them. Our friends the pedologists were lost in the middle of an indecipherable landscape (see Figure 2.7); once safely back in Manaus they had all the pedological horizons mapped out and could now master at a glance the forest that had previously dominated them. As can be seen in the frontispiece of the book by Mercator, the sixteenth-century geographer who first used the term *atlas*, the demiurgic task of Atlas—that of bearing the world on his shoulders—has now been transformed into "an atlas" and requires no more heroic force than that of turning the pages of a beautifully printed book that the cartographer holds in his hands.

This first loop deals with expeditions and surveys, with instruments and equipment, but also with the *sites* in which all the objects of the world thus mobilized are assembled and contained. For instance, here in Paris alone, the galleries of the Museum d'Histoire Naturelle, the collections of the Musée de l'Homme, the maps of the Service Géographique, the databases of the CNRS, the files of the police, and the equipment of the physiology laboratories of the Collège de France—all these are so many crucial objects of study for those who wish to understand the mediation through which humans, speaking to one another, increasingly speak truthfully about things. Thanks to a new survey and new data, an economist formerly without resources can start spitting out reliable statistics at the rate of thousands of columns per minute. An ecologist whom nobody used to take seriously can now intervene in a debate with beautiful satellite photographs that allow her, without budging from a Paris laboratory, to observe the advance of the forest in Boa Vista. A doctor, accustomed to treating clients case by case at the operating table, now has access to tables of symptoms based on hundreds of cases, provided by the hospital's record service.

If we want to understand why these people begin to speak more authoritatively and with more assurance, we have to follow this mobilization of the world, thanks to which things now present themselves in a form that renders them immediately useful in the arguments that scientists have with their colleagues. Through this mobilization the world is converted into arguments. To write the history of the first loop is to write the history of the transformation of the world into immutable and combinable mobiles*. In brief, it is the study of the writing of the "great book of nature" in characters legible to scientists, or, to put it another way, it is the study of the *logistics* that are so indispensable to the *logics* of science.

Autonomization

To convince someone, a scientist needs data (or more exactly *sublata*), but also someone to convince! The aim of the historians of the second part of the vascular system is to show how a researcher finds colleagues. I call this second loop *autonomization,* because it concerns the way in which a discipline, a profession, a clique, or an "invisible college*" becomes independent and forms its own criteria of evaluation and relevance. We always forget that specialists are produced from amateurs in the same way soldiers are made out of civilians. There have not always been scientists and researchers. It was necessary, with great effort, to extract chemists from alchemists, economists from jurists, sociologists from philosophers; or to obtain the subtle mixtures that produce biochemists out of biologists and chemists, social psychologists out of psychologists and sociologists. The conflict of disciplines is not a brake on the development of science, but one of its motors. The increase in the credibility of experiments, expeditions, and surveys presupposes a colleague capable of both criticizing and using them. What would be the use of obtaining ten million colored images from a satellite, if there were only two specialists in the world who could interpret them? An isolated specialist is a contradiction in terms. No one can specialize without the concurrent autonomization of a small group of peers. Even in the middle of the Amazon, our friends the soil scientists never stopped speaking in a virtual arena of colleagues with whom they were constantly arguing in absentia, as if the wooded landscape had been transformed into the wooden paneling of a conference room.

The analysis of scientific *professions* is certainly the easiest part of science studies and the one most easily understandable by scientists, who are never short of gossip on this topic. It deals with the history of associations and learned societies, as well as those of small cliques, groups, and clusters which form the seeds of all relationships among researchers. More generally, this analysis deals with the criteria by which one can distinguish, in the course of history, between a scientist and a virtuoso, an expert and an amateur, a central and a marginal researcher. How does one establish the values for a new profession, the meticulous control over titles and over barriers to entry? How does one impose a monopoly of competence, regulate the internal demography of a field, and find jobs for students and disciples? How does one resolve the innumerable conflicts of competence between the profession and its neighboring disciplines, between, say, botany and pedology?

In addition to the history of professions and disciplines, the second loop includes the history of scientific *institutions**. There must be organizations, resources, statutes, and regulations to keep the crowds of colleagues together. It isn't possible, for instance, to think of French science without the Académie, the Institut, the *grandes écoles*, the CNRS, the Bureau de Recherches Géologiques et Minières, and the Ponts et Chaussées. The institutions are as necessary for the resolution of controversies as is the regular flow of data obtained in the first loop. The problem for the practicing scientist is that the skills demanded for this second activity are entirely different from those of the first. A pedologist may be great at digging trenches and keeping worms in vats in the middle of the forest but utterly useless when it comes to writing papers and talking to colleagues. And yet one has to do both. Circulating reference does not stop with the data. It has to flow further and convince other colleagues as well. But things are even more complicated for scientists, because the circulation does not stop at this second loop either.

Alliances

No instruments can be developed, no discipline can become autonomous, no new institution can be founded without the third loop, which I call *alliances*. Groups that previously wouldn't give each other the time of day may be enrolled in the scientists' controversies. The military must be made interested in physics, industrialists in chemistry, kings in cartography, teachers in educational theory, congressmen in political science. Without this labor of making people interested, the other loops would be no more than armchair traveling; without colleagues and without a world, the researcher won't cost much but won't be worth much either. Immense groups, rich and well endowed, must be mobilized for scientific work to develop on any scale, for expeditions to multiply and go farther afield, for institutions to grow, for professions to develop, for professorial chairs and other positions to open up. The skills required for getting others interested are again different from those necessary for setting up instruments and for producing colleagues. One may be very good at writing convincing technical papers and terrible at persuading ministries that they cannot go on without science. As in the case of Joliot, these tasks can even be somewhat contradictory: his alliances bring in many strangers, like Dautry and his advisers, whereas the work of autonomization aims at limiting the discussion to his fellow physicists.

As we saw in the preceding section, it is not a question of historians finding a contextual explanation for a scientific discipline, but of the scientists themselves *placing the discipline in a context* sufficiently large and secure to enable it to exist and endure. It is not a matter of studying the impact of the economic base on the development of the scientific superstructure, but of finding out how, for example, an industrialist could improve his business by investing in a solid-state physics laboratory, or how a state geological service could expand by attaching itself to a department of transportation. The alliances do not pervert the pure flow of scientific information but are what makes this blood flow much faster and with a much higher pulse rate. Depending on the circumstances, these alliances can take innumerable forms, but this enormous labor of persuasion and liaison is never self-evident: there is no natural connection between a military man and a chemical molecule, between an industrialist and an electron; they do not encounter each other by following some natural inclination. This inclination, this *clinamen* has to be created, the social and material world has to be worked on to make these alliances appear, in retrospect, inevitable. This presents an immense and passionately interesting history, probably the most important for understanding our own societies: the history of how new nonhumans have become entangled in the existence of millions of new humans (see Chapter 6).

Public Representation

Even if the instruments were in place, if peers had been trained and disciplined, if well-endowed institutions were ready to offer a home to this wonderful world of colleagues and collections, and if government, industry, army, social security, and education provided the sciences with wide support, there would still be a great deal of work to be done. This massive socialization of novel objects—atoms, fossils, bombs, radar, statistics, theorems—into the collective, all this agitation, and all these controversies would present a terrible shock to people's everyday practice, would risk overturning the normal system of beliefs and opinions. It would be astonishing if it were otherwise, for what is science for if not to modify the associations of people and things? The same scientists who had to travel the world to make it mobile, to convince colleagues and lay siege to ministers and boards of directors, now have to take care of their relations with another outside world of civilians: reporters, pundits, and the man and woman in the street. I call this fourth loop *public representation* (if we can free this expression from the stigma associated with "PR").

Contrary to what is often suggested by science warriors, this new outside world is no more outside than the three previous ones: it simply has other properties and brings people with other qualities and competences into the fray. How have societies formed representations of what science is; what is a people's spontaneous epistemology? How much trust do they place in science? How can this confidence be measured in different periods and for different disciplines? How, for instance, was Isaac Newton's theory received in France? How was Charles Darwin's theory greeted by English clerics? How was Taylorism accepted by French trade-unionists during the Great War? How did economics little by little become one of the stock topics of politicians? How was psychoanalysis gradually absorbed into daily psychological discussions? How are DNA fingerprinting specialists faring on the witness stand?

Like all the others, this loop requires from scientists a completely different set of skills, unrelated to those of the other loops, and yet determinant for them all. One may be very good at convincing government ministers but completely unable to field questions on a talk show. How could one produce a discipline that would modify everyone's opinion, and nonetheless expect passive acceptance by all? If primatologists, ethologists, and geneticists produce entirely different genealogies for sex roles, aggression, and maternal love, how can they be surprised if large sections of the public take umbrage? Every astronomer recalculating the numbers of planets turning around stars knows that everything will change if throngs of other life forms are suddenly added to the definition of the human collective. This fourth loop is all the more important because the three others largely depend on it. A major part of advanced research in molecular biology in France, for example, depends on a private charity's annual telethon to combat muscular dystrophy. Every argument for or against genetic determinism will feed back into this funding. Our sensitivity to the public representation of science must be all the greater because information does not simply flow *from* the three other loops *to* the fourth, it also makes up a lot of the presuppositions of scientists themselves about their objects of study. Thus, far from being a marginal appendage of science, this loop too is part and parcel of the fabric of facts and cannot be left to educational theorists and students of media.

Links and Knots

To reach the fifth loop is not to reach the scientific content at last, as if the four others were simply conditions for its existence. From the first circle on, we have not departed for one moment from the course of scientific intelligence at work. As is clear from Figure 3.3, we have not been going endlessly around the mulberry bush and evading the "conceptual content," as science warriors are wont to say. We have simply followed the veins and arteries and arrive now, inevitably, at the pumping heart. Why does this fifth loop, which I call *links and knots* so as to avoid for the moment the word "concept," have the reputation of being much harder to study than the rest? Well, *it is* much harder. I don't pretend to crack it now, but simply to redefine its topology, so to speak, one of the reasons for its solidity.

This hardness is not that of a pit inside the soft flesh of a peach. It is that of a very tight knot at the center of a net. It is hard because it has to hold so many heterogeneous resources together. Of course, the heart is important for understanding the circulatory system of the human body, but Harvey certainly did not make his famous discovery by considering the heart on one side and the blood vessels on another. The same is true for science studies. If one takes the content on one side and the context on the other, the flow of science becomes incomprehensible, and so does the source of its oxygen and nutriment, as well as their means of entering the bloodstream. What would happen if there were no fifth loop? The other four would die off at once. The world would stop being mobilizable; disgruntled colleagues would flee in all directions; allies would lose interest; and so would the general public, after expressing either its shock or its indifference. But this death would ensue just as quickly if any of the other four loops were cut off.

This point is always one of the first casualties of the science wars. Of course Joliot "has thoughts"; of course he "has concepts"; of course his science has some content. But when science studies seeks to understand the centrality of the conceptual content of science, it tries first to see for *what* periphery this content plays the role of the center, of *what* veins and arteries it is the pumping heart, of *what* net it is the knot, of *what* pathways it is the intersection, of *what* commerce it is the clearing house. If we imagine Joliot to be now circulating along the loop that makes up the center of Figure 3.3, we understand why he tries so eagerly and so earnestly to find a way of keeping together, all at the same time, his instruments, his colleagues, the officials and industrialists he has gotten involved, and the public.

Yes, Joliot can succeed only by understanding the chain reaction—and he had better do it quick, before Szilard does it first, before the Germans arrive in Paris, before the two hundred liters of heavy water from the Norwegians run out, and before Halban and Kowarski are obliged to flee, denounced by their neighbors as foreigners. Yes, there is a theory, yes, the calculation of the cross-section made at night by Kowarski will make all the difference, yes, the knowledge they have produced about the neutrons will put them on the brink of a decisive advantage before the debacle of May 1940 brings an end to it all. But you need all the rest for this calculation to be the theory *of* something. There is indeed a conceptual core, but this is not defined by the preoccupations located at the *furthest remove* from the others; on the contrary, it is what keeps them all together, what strengthens their cohesion, what *accelerates their circulation*. Science warriors defend the conceptual content of science with the wrong sort of metaphor. They want it to be like an Idea floating in Heaven freed from the pollution of this base world; science studies wants to understand it as more like the heart beating at the center of a rich system of blood vessels, or better yet, like the thousands of alveoli in the lungs which allow the blood to be reoxygenated.

The difference in metaphors is not trivial. What science studies most wants to be able to explain is the relation between the *size* of this fifth loop and the four others. A concept does not become scientific because it is farther removed from the rest of what it holds, but because it is more intensely connected to a much larger repertoire of resources. Goat trails do not need turnpikes. Elephant hearts are a lot bigger than those of mice. The same goes for the conceptual content of a science: hard disciplines need bigger and harder concepts than soft ones, not because they are *more remote* from the rest of the world of data, colleagues, allies, and spectators—the four other loops—but because the world that they churn, steer, move, and connect is vastly bigger.

The content of a science is not something contained; it is itself a *container*. Indeed, if etymology is of any help, its *concepts*, its *Begriffe* (from *greifen*, to seize or to grasp), are what hold a collective tightly together. Technical contents are not astounding mysteries put in the way of those who study science by the gods to humble them by reminding them of the existence of another world, a world that escapes from history; nor are they provided for the amusement of epistemologists to enable them to look down on all those who are ignorant of science. They are part of this world. They only grow here, in our world, because they are what makes it up by linking together more and more elements in bigger and bigger collectives (as we will see in Chapter 6). For this point to be something more than an empty declaration of intent, I should obviously get much closer to the technical content than I have in my sketch of Joliot. But I cannot do this before substituting, in the next chapters, a new definition of what it is for a human to deal with a nonhuman for the old subject-object dichotomy. In the meantime, I can simply place concepts, links and knots, in a different position so that when we learn about the esoteric content of a science we immediately look for the four other loops that give it meaning.

The Enucleation of Society out of the Collective

How can I convince my scientific friends that by studying the vascularization of scientific facts we gain in realism and science gains in hardness? Perhaps this is so commonsensical that it appears heretical—for a little while at least. The more connected a science the sturdier it is; how could that be any simpler? And yet, for political reasons that will become clear in Chapter 7, epistemologists have transformed this very simple fact of life into a complete mystery. For epistemologists the scientific disciplines have to become solid and reliable without being connected through any sort of vessels to the rest of their world. The heart will be required to pump in and out, but there will be no input and no output, no body, no lungs, and no vascular system. Science warriors deal with nothing but an empty heart brightly lit on an operating table; science studies treats a bloody, throbbing, tangled mess, the entire vascularization of the collective. And the first group makes fun of the second because its members look messy and have blood on their white coats, and accuses them of ignoring the heart of science! Indeed, how can we talk to one another?

Yet, as at the end of Chapter 2, we also have to account for how the implausible, irrealist model can be extracted from the realist one proposed by science studies. A new paradigm should always be able to understand the one it claims to replace. As we saw in Figure 2.24, the notion of a yawning gap between words and world was obtained by erasing all the mediations and interrogating only the two extremes facing each other like two distant bookends, thus artificially creating the "problem" of reference. The mutilation of the circulatory system of science is even more gruesome (see Figure 3.4). If one fails to pay close attention to the entirety of scientific endeavor (Figure 3.4a) one can get the impression that there exists on the one hand a series of contingencies (the corona) and on the other hand, at the center, a conceptual content that counts most (Figure 3.4b). Here, it will take only the slightest lapse of attention, the slightest bit of carelessness, and that will be it! The rich and fragile webs will be cut, distanced from the things they connect and assemble. Another tiny slip, and the nucleus of "scientific content" will be separated from what will become, by contrast, a contingent historical "context" (Figure 3.4c). We will

(a)   (b)   (c)

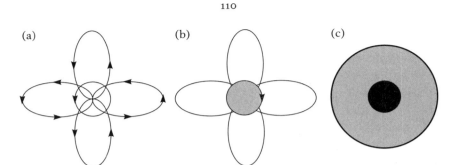

Figure 3.4   As in Figure 2.24, it is possible to extract the canonical model from the new one by erasing key mediations. If the conceptual dimension—the center circle in (a)—is excised from the other four, it will be transformed into a core (b), while the four other loops, now disconnected, when reconnected will form a sort of context of no relevance for defining the inner core of science (c).

have shifted from one branch of geometry to another, from knots to surfaces.

Only with inattention and the careless use of different analytical scalpels can one get the model of content vs. context from the heterogeneous and multiple labor of scientists. The whole of this labor then becomes obscure, because one no longer sees the essential connecting point, which is all the diverse elements that the theories and concepts theorize and bring together. Instead of the continuous and curved path of translations, one runs into an iron curtain separating the sciences from "extrascientific" factors, just as a long gray wall of concrete used to cut off the circulation through Berlin's delicate system of lanes, tramlines, and neighborhoods. Epistemologists, discouraged when faced with these objects so hard and so durable that they seemed to come from another world, could only send them to a Platonic Heaven and connect them to one another in an entirely phantasmagorical history, which is sometimes called the "conceptual history of science" despite the fact that there is no longer anything historical about it and *thus* nothing scientific about it either (see Chapter 5). The damage has been done: long trajectories of solid ideas and principles now appear to hover above a contingent history like so many foreign bodies.

The worst is yet to come: historians, economists, sociologists, accustomed to studying all the aspects I have listed, become discouraged by all this strangeness bobbing over their heads and leave the conceptual core of the sciences to the scientists and philosophers, modestly contenting themselves with wading through "social factors" and "social dimensions." This modesty would do them honor if, in abandoning the study of scientific and technical content, they did not also render incomprehensible *the very social existence* that they claim to study and to which they claim to restrict themselves. Indeed, what is most serious about this entirely artificial separation between the nucleus and the cell, of theories from what they theorize, is not that it enables intellectual historians to postulate this ahistorical, endless unfolding of "purely" scientific ideas. The real danger lies in the corresponding belief among social scientists that by lining up previously "enucleated" contexts it is possible to account for the existence of societies *without having to deal with science and technology.*

In place of a collective of humans and nonhumans we now have two parallel series of artifacts that never intersect: ideas on the one hand and *society** on the other. The first series, which results in the dreams of epistemology and the knee-jerk defensiveness of science warriors, is simply annoying and puerile; the second, which results *in the illusion of a social world,* is far more damaging, at least for those like me who try to practice a realistic philosophy. The whole of the modern world is made impossible to understand by this invention of an enucleated social context.

Let us suppose, for example, that a historian is studying the military decisions and programs of France during World War II. As we have seen, operations of translation made Joliot's laboratory indispensable to the conducting of French military affairs. Now, Joliot himself could not get his reactor to start except by discovering a new radioactive element, plutonium, which kicks off the chain reaction far more easily. Historians of military affairs, following the series of translations, must inevitably become interested in the history of plutonium; more precisely, this inevitability is a function of Joliot's work and his success. Given scientists' activities over the last three or four hundred years, how long can one study a military man before finding oneself in a laboratory? At most a quarter of an hour if one studies postwar science, and maybe an hour if one is dealing with the previous century (McNeill 1982; Alder 1997). Consequently, to write military history without looking at the laboratories that make up this history is an absurdity. It is not a matter of disciplinary principles, knowing whether or not one has the right to approach history without paying attention to science and technology; it is a question of *fact*—whether or not the players studied by historians mixed their lives and their feelings with nonhumans mobilized by laboratories and scientific professions. If the answer is yes, as it most certainly is in this example, it is unthinkable not to put back into the game the plutonium that Joliot and the military used, in their different ways, to make war and peace.

Now we can begin to measure the huge misunderstanding of those who say that science studies provides "a social explanation of science." Yes indeed, it offers an explanation, but of the *artifactual origin of a useless concept of society** that has been obtained by the enucleation of scientific disciplines out of their collective existence. What remains after this excision is, on the one hand, a society of humans among themselves and, on the other, a conceptual core. It would be even more absurd to say that science studies seeks to reconcile a social explanation with a conceptual explanation, if we understand these as the two distinct kinds of explanation that keep the parallel series of artifacts from ever intersecting. To retie two artifacts together makes for a third artifact, not for a solution! From Figure 3.4 it should be obvious that simply grafting a large corona of social factors onto the inner core of science, as in 3.4c, will not return us to the rich vascularization of scientific facts circulating through the five loops of 3.4a. The metaphors, the paradigms, the methods are entirely different and wholly incompatible. No matter how strange it may seem to science warriors and, yes, to most *social scientists,* in order to regain a sense of realism in the study of science, *one has to abandon the notion of a society altogether.* No wonder: as we will see in Chapters 7 and 8, this conception of society was invented for reasons entirely unsuited to providing an explanation of anything.

From Fabrication to Reality
Pasteur and His Lactic Acid Ferment

We have now made two moves that should begin to modify for good
the settlement* laid out in the first chapter. The notion of a world "out
there" to which a mind-in-a-vat tries to get access by establishing
some safe correspondence between words and states of affairs should
now be seen for what it is: a very unrealistic position for science, so
forced, so cramped that it can only be explained by powerful political
motives (which we will examine later in this book). In Chapter 2 we
began to understand that reference is not something that is added to
words, but that it is a circulating phenomenon, whose deambulation—
to borrow, once again, William James's term—should not be interrupted by any saltation if we want words to refer to the things progressively packed into them. Instead of the *vertical* abyss between words
and world, above which the perilous footbridge of correspondence
would hang, we now have a sturdy and thick layering of *transverse*
paths through which masses of transformations circulate.

Then in Chapter 3 we realized what an impossible double bind the
old settlement imposed on the scientist: "Be entirely cut off from the
weight of society, psychology, ideology, people"; and at the same time,
"Be absolutely, not relatively, sure of the laws of the world outside."
Against this contradictory injunction, we realized that the only reasonable, the only realistic way for a mind to speak truthfully about the
world is to *reconnect* through as many relations and vessels as possible
within the rich vascularization that makes science flow—and of course
this means that there is no longer any "mind" (Hutchins 1995). The
more relations a scientific discipline has, the more chance there is for

accuracy to circulate through its many vessels. Instead of the impossible task of freeing science from society, we now have a more manageable one: that of tying the discipline as much as possible to the rest of the collective.

And yet nothing is solved. We have simply begun to extract ourselves from the most blatant defects of the old settlement. We have not yet found a better one. *More reality* has to be taken into account if we want to continue. In Chapters 2 and 3 we left the world, so to speak, intact. Our friends the soil scientists, Joliot and his colleagues, were doing many things, but the soil itself, the neutrons themselves were behaving as if they had been there all along, waiting to be metamorphosed into so many stakes, diagrams, maps, arguments and brought to bear on the realm of human discourse. This is obviously not enough to explain how we can talk truthfully about a state of affairs. It does not matter how much we modify the notion of reference, if we are not also able to modify our understanding of what the entities of the world do when they come into contact with the scientific community and begin to be socialized into the collective*.

From the very beginning of science studies the solution has been to use the words "construction" and "fabrication." To take account of this transformation of the world effected by scientists, we have spoken of "the construction of facts," "the fabrication of neutrons," and other similar expressions which throw science warriors into fits and which they now fling back at us. I would be the first to admit that there are many problems with this way of accounting for action. First, although "construct" and " fabricate" are terms for technical activities, it happens that, under the pens of sociologists and philosophers working within the narrow space that the modern settlement allowed them, technology has been rendered almost as obscure as science (as we will see in Chapter 6). Second, this account implies that the initiative of action always comes from the human sphere, the world itself doing little more than offering a sort of playground for human ingenuity (in the discussion of the "factish" in Chapter 9 I will seek to counteract this). Third, speaking of construction implies a zero-sum game, with a fixed list of ingredients; fabrication merely combines them in other ways. Finally, and this is much more worrying, the old settlement has kidnapped the notions of construction and fabrication, turning them into weapons in a polarized battle against truth and reality. All too often the implication is that if something is fabricated it is false; likewise, if it is constructed it must also be deconstructible.

These are the main reasons why the more we in science studies showed the constructivist character of science, the deeper was the misunderstanding between us and our scientific friends. It is as if we were undermining science's claim to truth. Yes, we were undermining something, but something else altogether. Although we were a bit slow to realize it, we were shaking the foundations of the *very idiom of construction and fabrication* we had earlier taken for granted—and also, as we will see in Chapter 9, the basic notions of action and creation. Construction and fabrication, even more than reference and "conceptual content," have to be reconfigured totally, like all the other concepts that have been handed down to us, if we really wish to understand science in action. This reconfiguration is what I hope to accomplish in this chapter by visiting yet another empirical site, this time Louis Pasteur's laboratory. Let us follow in some detail the "Mémoire sur la fermentation appelée lactique,"[1] which historians of science consider to be one of Pasteur's most important papers.

The text is ideal for our purposes since it is structured around two combined dramas. The first one modifies the status of a nonhuman and that of a human. It converts a nonentity, the Cinderella of chemical theory, into a glorious and heroic character. In parallel, Pasteur's opinion, the Prince Charming, triumphs against all odds and reverses Liebig's theory: "The stone which the builders refused has become the cornerstone." And then there is a second drama, a reflexive one, a mystery that appears only at the end: who is constructing the facts, who is directing the story, who is pulling the strings? Is it the scientist's prejudices, or is it the nonhumans? Thus to the ontological drama is added an epistemological one. We will be able to see, using Pasteur's own words, how a scientist solves for himself and for us two of the basic problems of science studies. First let us turn to the uplifting story of Cinderella-the-yeast.

1. Partially translated into English by J. B. Conant in the Harvard Case Studies in Experimental Science (Conant 1957). I have completed and modified the translation in several places. The French text can be found in volume 2 of Pasteur's complete works. On the background see Geison (1974).

The First Drama: From Attributes to Substance

In 1856, some time after the fermentation of brewer's yeast became his primary interest, Pasteur related the discovery of a yeast peculiar to lactic acid. Today lactic fermentation is no longer an object of discussion, and dairies, creameries, and cheese manufacturers the world over can order by mail as much yeast as they need. But one has only to "place oneself in the conditions of the period" to appreciate the originality of Pasteur's report. In the middle of the nineteenth century, in scientific circles where Liebig's chemistry held sway, the claim that a specific microorganism could explain fermentation amounted to a step backward, since it was only by ridding itself of obscure vitalist explanations that chemistry had won its laurels. Fermentation had been explained in a purely chemical way, without the intervention of any living thing, by an appeal to the degradation of inert substances. In any case, specialists in lactic fermentation had never seen any microorganisms associated with the transformation of sugar.

At the beginning of Pasteur's paper, lactic acid fermentation has no clear-cut, isolable cause. If a yeast is involved it is nothing but an almost invisible by-product of a purely chemical mechanism of fermentation, or even worse, it is an unwelcome impurity that would hinder and spoil the fermentation. By the end of the paper, however, the yeast has become a full-blown entity in its own right, integrated into a class of similar phenomena; it has become the sole cause of fermentation. In the course of a single paragraph Pasteur takes the yeast through this entire transformation:

> Under the microscope, when one is not forewarned, it is *hardly possible to distinguish it* from casein, disaggregated gluten, etc.; in such a way that *nothing indicates that it is a separate* material or that it was produced during the fermentation. Its apparent weight *always remains very little* as compared to that of the nitrogenous material originally necessary for the carrying out of the process. Finally, very often it is *so mixed* with the mass of casein and chalk that there would be *no reason to suspect its existence.* (§7)

And yet Pasteur concludes this paragraph with this brave and surprising sentence: "It is *this* [the yeast], nevertheless, that plays *the principal role.*" The abrupt transformation is not only that of the yeast extracted from nothingness to become everything, it is also that of the Prince Charming, Pasteur himself. At the beginning of the paper, his opinion counts for nothing against Liebig's and Berzelius's powerful theories. At the end of the paper, Pasteur triumphs over his enemies and his view wins the day, defeating the chemical account of the fermentation. He begins:

The facts [that make the cause of lactic acid fermentation so obscure] then seem *very favorable to the ideas* of Liebig or to those of Berzelius . . . These opinions *gain more credit daily* . . . These *works all agree in rejecting the idea* of some sort of influence from organization and life as a cause of the phenomena that we are considering. (§5)

And again he concludes the paragraph with a defiant sentence that deflects the weight of previous arguments: "*I have been led to an entirely different point of view.*" But to accompany this elevation of Cinderella and this triumph of Prince Charming, another, more wideranging transformation is necessary. The capacities of the natural world are modified between the beginning and the end of the story. At the start of the paper the reader lives in a world in which the relation between organic matter and ferments is that of contact and decay:

In the eyes of [Liebig] *a ferment is an excessively alterable* substance that decomposes and thereby excites fermentation in consequence of its alteration which communicates a disintegrating disturbance to the molecular group of the fermentable matter. According to Liebig, such is the primary cause of all fermentations and the origin of most contagious diseases. Berzelius believes that the chemical act of fermentation is to be referred to the action of *contact*. (§5)

At the end the reader lives in a world in which a ferment is as active as any other already identified life form, so much so that it now feeds on the organic material, which instead of being its cause, has become its food:

Whoever judges impartially the results of this work and that which I shall shortly publish will recognize with me that fermentation appears to be correlative *to life* and to the *organization* of globules, and *not* to their *death* and putrefaction, no more than fermentation is a phenomenon due to contact in which the transformation of sugar takes place in the presence of the ferment without giving up anything to it or taking anything from it.(§22)

Let us now follow the main nonhuman character of the story to see through how many different ontological stages this entity is forced to pass before becoming something like a well-recognized substance. How does a scientist explain in his own words this emergence of a new actor out of other entities that he has to destroy, redistribute, and reassemble? What happens to this actant *x* that will soon be named lactic acid fermentation yeast? Like the forest-savannah limit in Chapter 2, the new entity is first a circulating object undergoing trials and submitted to an extraordinary series of transformations. At the beginning its very existence is denied:

> Until now minute researches have been *unable to discover the development of organized beings.* Observers who have recognized some of those beings have at the same time established that they were *accidental* and *spoiled* the process. (§4)

Then Pasteur's main experiment allows "a forewarned observer" to detect such an organized being. But this object *x* is stripped of all its essential qualities, which are redistributed among elementary sense data:

> If one carefully examines an ordinary lactic fermentation, there are cases where one can find, on top of the deposit of the chalk and nitrogenous material, *spots of a gray substance which sometimes form a layer [formant quelquefois zone]* on the surface of the deposit. At other times, this substance is found adhering to the upper sides of the vessel, where it has been carried by the movement of the gases. (§7)

> When it solidifies *[prise en masse]* it *looks* exactly *like* ordinary pressed or drained yeast. It is slightly *viscous,* and *gray* in color. Under the microscope, it appears to be formed of little *globules* or very short segmented filaments, isolated or in clusters, which form irregular flakes *resembling* those of certain amorphous precipitates. (§10)

It would be hard for something to have less existence than that! It is not an object but a cloud of transient perceptions, not yet the predicates of a coherent substance. In Pasteur's philosophy of science the phenomena precede what they are the phenomena of. Something else is necessary to grant *x* an essence, to make it into an actor: the series of laboratory trials through which the object *x* proves its mettle. In the next paragraph Pasteur turns it into what I have called elsewhere "a name of action*": we do not know what *it is*, but we know what *it does* from the trials conducted in the lab. A series of performances* *precedes* the definition of the competence* that will later be made the sole cause of these very performances.

> About fifty to one hundred grams of sugar are then dissolved in each liter, some chalk is added, and *a trace of the gray material* I have just mentioned from a good, ordinary lactic fermentation *is sprinkled* in . . . On the very next day a *lively and regular fermentation is manifest.* The liquid, originally very limpid, *becomes* turbid; little by little the chalk *disappears,* while at the same time a deposit *is formed* that grows continuously and progressively with the solution of the chalk. The gas that is *evolved* is pure carbonic acid, or a mixture in variable proportions of carbonic acid and hydrogen. After the chalk has *disappeared,* if the liquid is evaporated, an abundant crystallization of lactate of lime *forms* overnight, and the mother liquor contains variable quantities of the butyrate of this base. If the proportions of chalk and sugar are correct, the lactate *crystallizes* in a voluminous mass right in the liquid during the course of the operation. Sometimes the liquid *becomes* very viscous. In a word, we have before our eyes a *clearly characterized* lactic fermentation, with all the accidents and the usual complications of this phenomenon whose external manifestations are well known to chemists. (§8)

We do not yet know what it is, but we do know that it can be sprinkled, that it triggers fermentation, that it renders a liquid turbid, that it makes the chalk disappear, that it forms a deposit, that it generates gas, that it forms crystals, that it becomes viscous (Hacking 1983). As of now it is a list of entries recorded in the laboratory notebook, *membra disjecta* which do not yet pertain to one entity—properties looking for the substance they belong to. At this point in the text, the entity is so fragile, its *envelope** so indeterminate, that Pasteur notes with surprise its ability to travel:

> It *can* be collected and *transported* for great distances without losing its activity, which is *weakened* only when the material is dried

or when it is boiled in water. Very little of this yeast is necessary to transform a considerable weight of sugar. These fermentations should *preferably* be carried on so that the material is protected from the air, so that they will not be hindered by vegetation or foreign infusoria. (§10)

Maybe shaking the flask will make the phenomenon disappear, maybe exposing it to the air will destroy it. Before the entity is safely underwritten by a fixed ontological substance, Pasteur has to add precautions that he will soon find useless. Not yet knowing what it is, he has to fumble, investigating all sides of the vague boundaries he has sketched around the entity in order to determine its precise contours.

But how can he increase the ontological status of this entity, how can he transform these fragile, uncertain boundaries into a sturdy envelope, how can he move from this "name of action" to the "name of a thing"? If it acts so much, must the entity be an actor? Not necessarily. *Something more* is needed to turn this fragile candidate into a full-blown actor which will be designated as the origin of those actions; another act is necessary to conjure up the substrate of these predicates, to define a competence that will then be "expressed" or "manifested" through so many performances in laboratory trials. In the main section of the paper, Pasteur does not hesitate. He uses everything at hand to stabilize the noumenal substrate of this entity, granting it an activity similar to that of brewer's yeast. Borrowing the metaphor of growing plants allows him to evoke the processes of domestication and cultivation, the firmly established ontological status of plants, as a way of giving shape to his aspiring actor:

> Here we find *all the general characteristics* of brewer's yeast, and these substances probably have organic structures that, in a natural classification, place them in *neighboring species* or in two connected families. (§11)

> There is another characteristic that permits one to compare this new ferment with brewer's yeast: if brewer's yeast instead of the lactic ferment *is sown* in limpid, sugared, albuminous liquid, brewer's yeast will develop, and with it, alcoholic fermentation, even though the other conditions of the operation remain unchanged. One should not conclude from this that the chemical composition of the two yeasts is identical any more than that the chemical composition of *two plants* is the same because they grew *in the same soil*. (§13)

What was a nonentity in §7 has become so well established in §11 that it has a name and a place in the most precise and most venerable of all branches of natural history, taxonomy. No sooner has Pasteur shifted the origin of all the actions to the yeast, which thereby becomes a full-blown independent entity, than he uses it as a stable element to redefine all the former practices: we did not know what we were doing before, but now we do:

All the chemists will be surprised at the rapidity and regularity of lactic fermentation under the conditions that I have specified, that is, *when the lactic ferment develops alone;* it is often more rapid than the alcoholic fermentation of the same amount of material. Lactic fermentation *as it is ordinarily carried out* takes much longer. This can easily be *understood.* The gluten, the casein, the fibrin, the membranes, the tissues that are used contain an enormous amount of useless matter. More often than not these become a *nutrient* for the lactic ferment only after putrefaction—alteration by contact with plant or animalcules—that has rendered the elements soluble and assimilable. (§12)

A slow and uncertain practice with an obscure explanation becomes a quick and comprehensible set of new methods mastered by Pasteur: all along, without knowing it, cheese manufacturers had been cultivating microorganisms in a medium that provides food for the ferment, food that itself may be varied so as to vary the adaptation to an environment of multiple ferments in competition. What was the primary cause of a useless by-product has been transformed into food for its consequence!

Going even further, Pasteur turns this newly shaped entity into one "singular case" within a whole class of phenomena. The "general circumstances" of this widespread phenomenon, fermentation, can now be defined.

One of the essential conditions for *good fermentations* is the *purity* of the ferment, its *homogeneity,* its *free development without any hindrance* and with the help of a nutrient well *adapted* to its individual nature. In this respect, it is important to realize that the *circumstances* of neutrality, of alkalinity, of acidity, or of the chemical composition of the liquids play an important part in the predominant growth of such and such a ferment, because the life of each does not *adapt* itself to the same degree to different states of the *environment.* (§17)

By drawing on several seemingly incompatible philosophies of science, Pasteur provides a fresh solution to what is still a subject of much controversy in epistemology, namely, how a new entity can emerge out of an old one. It is possible to go from a nonexistent entity to a generic class by passing through stages in which the entity is made of floating sense data, taken as a name of action, and then, finally, turned into a plantlike and organized being with a place within a well-established taxonomy. The circulation of reference does not take us, as in Chapters 2 and 3, from one site of research to the next, from one type of trace to the next, but *from one ontological status to the next.* Here it is no longer just the human who transports information through transformation, but the nonhuman as well, surreptitiously changing from barely existing attributes into a full-blown substance.

From Fabrication of Facts to Events

How does Pasteur's own account of the first drama of his text modify the commonsense understanding of fabrication? Let us say that in his laboratory in Lille Pasteur is *designing* an *actor.* How does he do this? One now traditional way to account for this feat is to say that Pasteur designs trials* for the actor* to show its mettle. Why is an actor defined through trials? Because there is no other way to define an actor but through its action, and there is no other way to define an action but by asking what other actors are modified, transformed, perturbed, or created by the character that is the focus of attention. This is a pragmatist tenet, which we can extend to (a) the thing itself, soon to be called a "ferment"; (b) the story told by Pasteur to his colleagues at the Academy of Science; and (c) the reactions of Pasteur's interlocutors to what is so far only a story found in a written text. Pasteur is engaged at once in *three* trials that should be first *distinguished* and then *aligned* with one another, according to the notion of circulating reference, with which we are now familiar.

First, in the story told by Pasteur, there are characters whose competence* is defined by the performances* they undertake: the nearly invisible Cinderella becomes, to the applause of the reader, the hero who triumphs and becomes the essential cause of lactic fermentation, of which it was, at first, a useless by-product. Second, Pasteur, in his laboratory, is busy staging a new artificial world in which to try out this new actor. He does not know what the essence of a ferment is. Pasteur is a good pragmatist: for him essence is existence and existence is action. What is this mysterious candidate, the ferment, up to? Most of an experimenter's ingenuity goes into designing devious plots and careful staging that make an actant* participate in new and unexpected situations that will actively define it. The first trial is a story: it pertains to language and is similar to any trial in fairy tales or myths. The second is a situation: it pertains to nonverbal, nonlinguistic components (glassware, yeasts, Pasteur, laboratory assistants). Or does it?

The third trial is designed *to answer this very question.* Pasteur undergoes this new trial when he tells his story of the Cinderella that triumphs against all odds and of the Prince Charming that defeats the dragon of chemical theory—when he has a shorter version of his paper read at a meeting of the Academy on 30 November 1857. Pasteur is now trying to convince the Academicians that his story is not a story, but that it has occurred *independently* of his wishes and imaginative ability. To be sure, the laboratory setting is artificial and manmade, but Pasteur must establish that the competence of the ferment is *its* competence, *in no way* dependent on his cleverness in inventing a trial that allows it to reveal itself. What happens if Pasteur wins this new (third) trial? A new competence will now be added to *his* definition. Pasteur will be the person who has shown, to everyone's satisfaction, that yeast is a living organism, just as the second trial added a new competence to this other actant, the ferment: namely that it can trigger a specific lactic fermentation. What happens if Pasteur fails? Well, the second trial will have been a waste. Pasteur will have entertained his peers with the tale of Cinderella-the-ferment, an amusing story to be sure, but one which will have involved his own expectations and earlier prowess only. Nothing new will have been conveyed by Pasteur's words at the Academy to modify what his colleagues say about him and about the abilities of living organisms that make up the world.

However, an experiment is none of these three trials in isolation. It is the *movement* of the three *taken together when it succeeds, or separated when it fails.* Here we recognize again the movement of circulating reference we studied in Chapter 2. The accuracy of the statement is not related to a state of affairs out there, but to the traceability of a series of transformations. No experiment can be studied only in the laboratory, only in the literature, or only in the debates among colleagues. An experiment is a story, to be sure—and studiable as such—but a story *tied* to a situation in which new actants undergo terrible trials plotted by an ingenious stage manager; and then the stage manager, in turn, undergoes terrible trials at the hands of his colleagues, who test what sort of *ties* there are between the first story and the second situation. An experiment is a text about a nontextual situation, later tested by others to decide whether or not it is simply a text. If the final trial is successful, then *it is* not just a text, there is indeed a real situation *behind* it, and both the actor and its authors are endowed with a new competence: Pasteur has proved that the ferment is a living thing; the ferment is able to trigger a specific fermentation different from that of brewer's yeast.

The essential point I am trying to make is that "construction" is in no way the mere recombination of already existing elements. In the course of the experiment Pasteur and the ferment *mutually exchange and enhance their properties,* Pasteur helping the ferment show its mettle, the ferment "helping" Pasteur win one of his many medals. If the final trial is lost, then it was just a text, there was nothing behind it to support it, and neither actor nor stage manager has won any *additional* competences. Their properties cancel each other out, and colleagues can conclude that Pasteur has simply prompted the ferment to say what he wished it to say. If Pasteur wins we will find two (partially) new actors on the bottom line: a new yeast and a new Pasteur! If he loses, there will be only one, and he, the Pasteur of old, will go down in history as a minor figure together with a few shapeless yeasts and wasted chemicals.

We need to understand that whatever we want to think or argue about the artificial character of the laboratory, or the literary aspects of this peculiar type of exegesis, the lactic acid ferment is invented *not* by Pasteur but *by the ferment.* At least, this is the problem that the trials of his colleagues, of Pasteur himself, and of the little bug in the glassware must resolve. It is essential to all of them that whatever the ingenuity of the experiment, whatever the perverse artificiality of the setup, whatever the underdetermination or the weight of theoretical expectations, Pasteur manages to take himself out of the action so as to become an expert, that is, *experitus,* someone transformed by the manifestation of something not contrived by the former Pasteur. No matter how artificial the setting, something new, independent of the setting, has to emerge, or else the whole enterprise is wasted.

It is because of this "dialectic" between fact and artifact that, although no philosopher would seriously defend a correspondence theory of truth, it is nevertheless absolutely impossible to be convinced by a purely constructivist argument for more than three minutes. Well, let's say an hour, to be fair. Most philosophy of science since Hume and Kant consists in taking on, evading, hedging, coming back to, recanting, solving, refuting, packing, unpacking this impossible antinomy: that on the one hand facts are experimentally made up and never escape from their manmade settings, and on the other hand it is essential that facts are *not* made up and that something emerges that is *not* manmade. Bears in cages pace back and forth within their narrow prisons with less obstinacy and less distress than philosophers and sociologists of science going incessantly from fact to artifact and back.

This obstinacy and this distress come from defining an experiment as a zero-sum game. If the experiment is a zero-sum game, if every output has to be matched by an input, then nothing escapes from a laboratory that has not been previously put into it. Such is the real weakness of common definitions of construction and fabrication: whatever the philosopher's list of the inputs in a setting, it always features the *same* elements before and after—the same Pasteur, the same ferment, the same colleagues, or the same theory. Whatever the scientists' genius, they always play with a fixed set of Lego blocks. Unfortunately, since it is at once fabricated and not fabricated, there is always *more* in the experiment than was put into it. Explaining the outcome of the experiment by using a list of stable factors and actors will therefore always show a *deficit*.

It is this deficit that will then be accounted for differently by the various realist, constructivist, idealist, rationalist, or dialectic persuasions. Each will *make up* the deficit by cashing in its favorite stocks: nature "out there," macro- or micro-social factors, the transcendent Ego, theories, standpoints, paradigms, biases, or the churning blender of dialecticians. There seems to be an endless supply of fat bank accounts upon which one can draw to complete the list and "explain" away the originality of an experiment's outcome. In this kind of solution, the novelty is not accounted for by modifications in the list of initial actors, but by the addition of one paramount factor that *balances* the account. In this way, every input is balanced by an output. Nothing new has happened. Either experiments simply reveal Nature; or alternatively society, or biases, or theoretical blind spots betray themselves in the outcome, over the course of an experiment. Nothing more happens in the history of science than the discovery of what was already there, all along, in nature or in society.

But there is no reason to believe that an experiment is a zero-sum game. On the contrary, each of the difficulties posed by Pasteur's paper suggests that *an experiment is an event**. No event can be accounted for by a list of the elements that entered the situation *before* its conclusion, *before* Pasteur launched his experiment, *before* the yeast started to trigger the fermentation, *before* the meeting of the Academy. If such a list were made, the actors on it would not be endowed with the competence that they will *acquire* in the event. On this list Pasteur is a promising crystallographer but has not shown to anyone's satisfaction that the ferments are living creatures; the ferment may accompany the fermentation, as Liebig allowed, but it is not yet endowed with the property of triggering a lactic acid fermentation different from that of brewer's yeast; as for the Academicians, they do not yet depend on a living yeast in their own laboratories and may prefer to remain on the solid foundations of chemistry they learned from Liebig instead of flirting again with vitalism. This list of inputs does not have to be completed by drawing upon any stock of resources, since the stock drawn upon *before* the experimental event is not the same as the one drawn upon *after* it. This is precisely why an experiment is an event and not a discovery, not an uncovering, not an imposition, not a synthetic *a priori* judgment*, not the actualization of a potentiality*, and so on.

This is also why the list drawn up after the experiment needs no addition of Nature, or society, or whatever, since all the elements have been partially transformed: a (partially) new Pasteur, a (partially) new yeast, and a (partially) new Academy are all congratulating one another at its end. The ingredients on the first list are insufficient, not because one factor has been forgotten or because the list has not been carefully drawn, but because actors *gain* in their definitions through this event, through the very trials of the experiment. Everyone agrees that science grows through experiment; the point is that Pasteur also is modified and grows through this experiment, as does the Academy, and, yes, the yeast too. They *all* leave their meeting in a different state from the one in which they entered. As we will see in the next chapter, this may lead us to inquire whether there is a history of science, not only of scientists, and whether there is a *history of things*, not only of science.

The Second Drama: Pasteur's Solution to the Conflict between Constructivism and Realism

If it has not been too difficult to reconfigure the notion of construction and fabrication so as to consider an experiment as an event, and not as a zero-sum game, it is much trickier to understand how we can simultaneously insist on the artificiality of the laboratory setting and also on the autonomy of the entity "made up" inside the laboratory walls. To be sure, we are helped by the double meaning of the word "fact"— that which is made and that which is not made up; "un fait est fait," as Gaston Bachelard put it—but a lot of conceptual work is necessary to probe the hidden wisdom of this etymology (see Chapter 9). It is easy to understand why houses and cars and baskets and mugs are at once *fabricated* and *real*, but this is of no help in accounting for the mystery of scientific objects. It is not just that they are both made up *and* real. Rather, it is precisely *because* they have been artificially made up that they gain a complete autonomy from any sort of production, construction, or fabrication. Technical or industrial metaphors are not going to help us grasp this most puzzling phenomenon, which has taxed the patience of science studies for so many years. As I have often found to be the case, the only solution when faced with difficult philosophical questions is to dive even deeper into some empirical sites to see how scientists themselves get out of the difficulty. Pasteur's solution in this paper is so clever that if we had followed it all along science studies would have taken an entirely different course.

Pasteur is perfectly aware that there is a gap in his genealogy. How can he go from the barely visible, gray matter that sometimes appears on the top of the vessel to the plantlike, full-blown substance endowed with nutritional needs and rather particular tastes? How can he make this crucial step? Who is responsible for the attribution of these actions, and who is responsible for the endowment of properties? Is Pasteur not giving his entity a little nudge forward? Yes, *he is* doing the action, he has prejudices, he fills the gap between underdetermined facts and what should be visible. He "confesses" it very explicitly in the very last paragraph of the paper:

> All through this memoir, I have reasoned *on the basis of the hypothesis* that the new yeast is organized, that *it is* a living organism, and that its chemical action on sugar *corresponds* to its development and organization. If someone were to tell me that in these conclusions I am going *beyond that which the facts prove*, I would answer that it is quite true, in the sense that the stand I am taking is in a framework of ideas *[un ordre d'idées]* that in rigorous terms *cannot be irrefutably demonstrated.* Here is the way I see it. Whenever a chemist makes a study of these mysterious phenomena and has the good fortune to bring about an important development, he will instinctively be *inclined* to assign their primary cause to a type of reaction *consistent* with the general results of his own research. It is the *logical* course of the human mind in all controversial questions. (§22)

Not only does Pasteur develop a whole ontology in order to follow the transformation of a nonentity into an entity, as we saw in the last section, but he also has an epistemology, and a pretty sophisticated one at that. Like most French scientists, he is a constructivist of the rationalist kind—against the positivism of his *bête noire*, Auguste Comte. For Pasteur facts always need to be framed and built up by a theory. The origin of this inevitable *"ordre d'idées"* is to be found in disciplinary loyalties ("a chemist"), themselves tied to past investment (*"consistent* with the general results of his own research"). Pasteur roots this disciplinary inertia in culture and personal history ("his own research") as well as in human nature ("instinct," "the logical course of the human mind"). In his own eyes, does the confession of this prejudice weaken Pasteur's claims? Not a bit—and this is the apparent paradox that is so important for us to understand. The very next sentence, which I have already quoted, introduces another quite different epistemology, a much more classical one in which facts may be unambiguously evaluated by impartial observers. In the remainder of this chapter we will try to understand this gap between two opposing sentences which, curiously, are not taken as contradictory.

And it is my opinion, at this point in the development of my knowl-edge of the subject, that *whoever judges impartially* the results of this work and that which I shall shortly publish *will recognize with me* that fermentation appears to be correlative to life and to the organization of globules, and not to their death and putrefaction. (§22)

Whereas in the sentence just before this one the logical course of the human mind precluded "impartial judgment," especially in "contro-versial questions" which cannot be "irrefutably demonstrated," it is suddenly possible for the same Pasteur to convince anyone who judges impartially. *Two entirely unrelated epistemologies are juxtaposed* without the slightest suggestion that there may be some difficulty here. First, facts need a theory if they are to be made visible, and this theory is rooted in the previous history of the research program—it is "path de-pendent" as economists would say—but then, facts may be judged in-dependently of earlier history. Once again the mystery of the two opposed meanings of the little word "fact" is reiterated. Is Pasteur un-aware of the difficulty, or are we unable to reconcile constructivism with empiricism as readily as he does? Whose contradiction is this— Pasteur's or ours?

In order to grasp how Pasteur, *without giving any sign of being para-doxical,* can go from one epistemology to its polar opposite, we have to understand how he distributes activity between himself, as the experi-menter, and the would-be ferment. An experiment, as we just saw, is an action performed by the scientist so that the nonhuman can be made to appear on its own. The artificiality of the laboratory does not run counter to its validity and truth; its obvious immanence is actually the source of its downright transcendence. How could this apparent miracle be obtained? Through a very simple setup that has baffled ob-servers for a long time and that Pasteur beautifully illustrates. The ex-periment creates two planes: one in which the narrator is active, and a second in which the action is delegated to another character, a nonhu-man one (see Figure 4.1).

An experiment *shifts out** action from one frame of reference to an-other. Who is the active force in this experiment? Both Pasteur *and* his yeast. More precisely, Pasteur acts *so that* the yeast acts alone. We understand why it is difficult for Pasteur to choose between a constructivist epistemology and a realist one. Pasteur creates a stage in

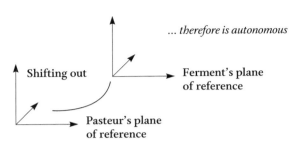

The ferment is constructed by Pasteur's hand and ...

Figure 4.1 The difficulty of accounting for an experiment comes from the "shifting-out" that relates the scientist's plane of reference to the object's plane of reference. It is only because Pasteur has worked well and hard in his own plane that the ferment is allowed to live autonomously in its own plane. This crucial connection should not be broken.

which he does not have to create anything. He develops gestures, glassware, protocols, so that the entity, once shifted out, becomes independent and autonomous. According to which of these two contradictory features is stressed, the same text becomes either constructivist or realist. Am I, Pasteur, making up this entity because I am projecting my prejudices onto it, or am I being made up and forced to behave that way because of *its* properties? Am I, the analyst of Pasteur, explaining the closure of the controversy by appealing to his human, cultural, historical interests, or will I be forced to add to the balance the active role of the nonhumans he did so much to shape? These questions are not philosophical problems confined to the pages of journals in the philosophy of science or the pitiful stakes of the science wars: they are the very questions tackled over and over by scientific papers, and by which they sink or swim.

The experimental scenography in Pasteur's paper is extremely varied since it follows all the subtleties of the variable ontology deployed in the text. In the same paper some experiments are backgrounded and blackboxed while others are made the focus of attention and are allowed to go through changes. At first the practice of doing science is alluded to only through very stylized accounts of experiments which are quickly blackboxed. In another case, human agency is reintroduced in a recipe-like description of the procedure that leads to lactic acid fermentation. But at this point there is no "trouble with experiments," to use Shapin and Schaffer's expression (Shapin and Schaffer 1985). The fermentation of lactic acid is a well-known procedure which Pasteur imports unchanged. He states, "Lactic acid was discovered by Sheele in 1780 in soured whey. His procedure for removing it from the whey is still today the best one can follow" (§4); he then includes the recipe. Firmly tied to practice but completely relegated to the background, this experimental procedure defines the baseline—lactic fermentation—out of which the foregrounded yeast will be made to appear. Without a stabilized recipe for lactic fermentation no yeast could start to show its mettle. In a single scientific paper the author may go through several philosophies of experiment with relativist or constructivist moments preceded by brutal denials of the role of instruments and human interventions and followed by positivist declarations. Pasteur's scenography, for instance, changes completely in the central paragraphs 7 and 8, in which the main experiment is displayed. Human activity is back under the spotlight, and so are the troubles that come with it:

> *I extract* the soluble part from brewer's yeast, by *treating* the yeast for some time with fifteen to twenty times its weight of water at the temperature of boiling water. The liquid, a complex solution of albuminous and mineral material, *is carefully filtered*. About fifty to one hundred grams of sugar are then *dissolved* in each liter, some chalk is *added,* and a trace of the gray material I have just mentioned from a good, ordinary lactic fermentation is *sprinkled* in; then one *raises* the temperature to 30 or 35 degrees centigrade. It is also *good* to *introduce* a current of carbonic acid in order to expel the air from the flask, which is *fitted* with a bent exit tube immersed under water. On the very next day a lively and regular fermentation is manifest . . . In a word, we have before our eyes a clearly characterized lactic fermentation, *with all the accidents and the usual complications* of this phenomenon whose external manifestations are well known to chemists. (§8)

At the very moment when the entity is at its weakest ontological status (see the first section of this chapter), shuffled among clouds of chaotic sense data, the experimental chemist is *in full activity*, extracting, treating, filtering, dissolving, adding, sprinkling, raising the temperature, introducing carbonic acid, fitting tubes, and so on. But then, shifting the attention of the reader, shifting out the autonomous actor, Pasteur says that "we have before our eyes a clearly characterized lactic fermentation." The director withdraws from the scene, and the reader, merging her eyes with those of the stage manager, *sees* a fermentation that takes form at center stage *independently* of any work or construction.

Who is doing the action in this new medium of culture? *Pasteur,* since he sprinkles, boils, filters, and sees. *The lactic acid yeast,* since it grows fast, uses up its food, gains power ("very little of this yeast is necessary to transform a considerable weight of sugar"), and enters into competition with other similar beings growing like plants on the same plot of land. If we ignore Pasteur's work, we slip into the pit of naive realism from which twenty-five years of science studies have tried to extract us. But what happens if we ignore the lactic acid's delegated automatic autonomous activity? We fall into the other pit, as bottomless as the first, of social constructivism, ignoring the role of nonhumans, on whom all of the people we study are focusing their attention, and for whom Pasteur spent months of labor designing this scenography.

We cannot even claim that in both cases it is only the author, the human author, who is doing the work in the writing of the paper, since what is at stake in the text is precisely the reversal of authorship and authority: *Pasteur authorizes the yeast to authorize him to speak in its name.* Who is the author of the whole process and who is the authority in the text are themselves open questions, since the characters and the authors exchange credibilities. As we saw in the previous section, if his colleagues at the Academy do not believe Pasteur, he will be made the sole and only author of a work of *fiction.* If the whole setup withstands the Academy's scrutiny, then the text itself will be in the end authorized by the yeast, the real behavior of which can then be said to *underwrite* the entire text.

How can we understand the artificial stagecraft of the experiment that aimed at letting the lactic acid develop alone, by its own agency, in a pure medium of culture? Why is it so complicated to recognize that an experiment is precisely the place where this contradiction is staged and resolved? Pasteur is not plagued here by false consciousness, erasing the traces of his own work as he goes along. We do not have to choose between two accounts of scientific work, since he explicitly places both of these two contradictory requirements in the final paragraph of the paper. "Yes," he says, "I went well beyond the facts, I had to, but any impartial observer will recognize that lactic acid is made of living organisms and not of dead chemistry." Acknowledging his activity does not, in his view, weaken his claim for the independence of the yeast, any more than seeing the threads in a puppeteer's hands weakens the credibility of the story enacted by the puppets "freely" acting in the other plane of reference. As long as we do not understand why what appears to us as a contradiction is not one for Pasteur, we fail to learn from those we study—we simply impose our philosophical categories and conceptual metaphors on their work.

In Search of a Figure of Speech: Articulation and Proposition

Is it possible to use these categories and figures of speech (even if it means reconfiguring them again), not to obscure the scientists' work, but to make it both visible and capable of producing results that are independent of it? Science studies has struggled so much with this question to no avail: why tackle it again? It would be much easier, I agree, to stick with the older settlement and accept the results of the philosophy of language, without bothering to attempt to engage the world with what we say about it, an attempt that seems to force us into so many intractable metaphysical difficulties. Why not go back to philosophical common sense, and simply distinguish epistemological questions from ontological ones? Why not limit history to people and societies, leaving nature immune to history altogether? Does science studies, to be understood, really require so much philosophical work (conceptual bricolage would be a more fitting name for it)? Why not rest quietly at some happy medium and say for instance that our knowledge is the *resultant* of two contradictory forces—to use the parallelogram of forces we all learned in primary school and David Bloor's version of it as taught in Science Studies 101 (Bloor [1976] 1991)? Everyone would be happy. We would have the power of societies, biases, paradigms, human feelings on the one hand, and those of nature and reality on the other, knowledge being simply the resulting diagonal. Would that not solve all the difficulties (see Figure 4.2)?

Unfortunately, there is no going back to nosh on the onions of Egypt

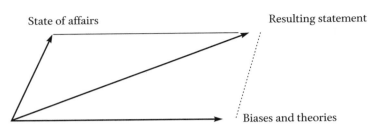

Figure 4.2 One classical solution to the problem of experiment is to consider it as the resultant of two forces, one representing the contribution of the empirical world and the other the contribution of a given system of beliefs.

that the Hebrews found, in retrospect, so palatable. The safe haven of the modern settlement is nostalgia, a form of exoticism (see Chapter 9); nothing really worked in that impossibly makeshift arrangement of contradictory positions. It is only because we are used to what we left behind and not to what we face now that we find the old settlement more commonsensical. How unreasonable this reasonable compromise really is.

According to the physics of the parallelogram, if no force at all came from the axis that I call "biases and theories," we would have a direct, pristine, unfettered access to a state of affairs. What laboratory scientist would believe that for a minute? Not Pasteur, at any rate, who knows well enough the work he puts into making a state of affairs visible, and knows that this work is what gives an accurate reference to the paper he presents to his Academy colleagues. But the opposite position, imputed to science studies by the science warriors, is even more implausible. If there were no pull at all from the axis I call "state of affairs," our statements about the world would be entirely made up of nothing but the *earlier* repertoire of myths, theories, paradigms, biases that society has in stock. What laboratory scientist could believe that for one minute—or what science student, for that matter? Not Pasteur in any case.

Where, in the repertoire and social prejudices of the nineteenth century, would one find anything to make up, to conjure, to slap together a little bug like the lactic acid ferment in Pasteur's flasks? No imagination is fertile enough for that feat of fiction. Surely a tug-of-war between two contrary forces will not do the job. No, no, the modern settlement works as long as one does not think too much about it and applies it unreflexively by shifting between completely contradictory positions. Only an enormously powerful political reason—see Chapters 7 and 8—can explain why we attribute the label of common sense to such an unrealistic definition of what it is to speak truly about a state of affairs. We may be uneasy about quitting our old habits of thought, but no one can say that we are abandoning reasonable positions for extravagant claims. If anything, in spite of the furious volleys of the science wars, we may be slowly moving from absurdity to common sense.

The difficulty of understanding Pasteur's solution comes from his using the two statements "the ferment has been fabricated in my laboratory" and "the ferment is autonomous from my fabrication" as *synonyms*. More precisely, it is as if he were saying that *because* of his careful and skillful work in the laboratory, the ferment is *therefore* autonomous, real, and independent of any work he has done. Why do we have so much trouble accepting this solution as common sense, and why do we feel obliged to protect Pasteur from committing one of the two analytical crimes? Either forgetting the work he has been doing so he can say that the ferment is "out there," or else abandoning the notions of nonhumans out there, so as to be able to focus our attention on his work? To illuminate what happens in an experiment, the metaphor of the parallelogram of forces leaves much to be desired. What other figures of speech might be better aids for understanding Pasteur's curious brand of what could be called "constructivist realism"?

Let's begin with the metaphor of *staging*, which I used in the previous section, with Pasteur as the director bringing certain aspects of the experiment to the foreground and backgrounding others outside the spotlights' glow. This metaphor has the great advantage of focusing attention on the two planes of reference at once, instead of making them pull in opposite directions. Although the work of the stage manager—or that of a puppeteer—clearly aims at its own disappearance, directing attention away from what happens backstage and toward what happens on the boards, it is clearly indispensable for the performance to take place. Most of the pleasure of the audience actually comes from the trembling presence of this other plane which is at once constantly felt and happily forgotten. However, with this pleasure comes the main weakness of this figure of speech. This metaphor, borrowed from the world of art, has the unfortunate consequence of *aestheticizing* the work of science and weakening its claim to truth. Although it may be accurate to say that a major effect of science studies has been to render the sciences pleasurable (Jones and Galison 1998), we are not looking for pleasure but for a truth independent of our own making.

Comparing science with art is of course less damaging than understanding science by using the notion of fetishism*, which we will study in Chapter 9. When scientists are portrayed as fetishists, they are accused of forgetting entirely the work they have just done and of being taken in by the seeming autonomy of the product of their own hands. Artists, at least, can enjoy the quality of the labor even when it vanishes from view, but there is nothing to redeem naive believers who forget that they are the sole cause of statements they believe to have no other cause than a thing out there. To be sure, this figure of speech accounts well for the forced disappearance of any telltale trace of labor, but, alas, it puts the laborers into a perverse position: scientists are seen either as clever manipulators of ventriloquous phenomena or as credulous magicians surprised by their own sleight of hand. We are not yet in a position to resolve this difficulty, which arises from the fundamental definitions of action and creation used by the modernists—this will have to wait until later, when I will introduce the strange concept of factish*. Can we do better and escape from art and make-believe altogether?

Why do I portray Pasteur as someone who "gazes" at the lactic acid ferment? Why do I use optical metaphors of *seeing?* The advantage of this way of speaking is that, although it does not capture in any way the activity of the one who looks, it does emphasize the independence and autonomy of the thing to be looked at. The optical metaphor is used endlessly by those who say that scientists have "tinted lenses" that "filter" what they "see," that they have "biases" "distorting" their "vision" of an object, that they have "world views" or "paradigms" or "representations" or "categories" with which they "interpret" what the world is like. With such expressions, however, it is utterly impossible for these mediations to be anything but *negative,* since, in contrast to these expressions, the ideal of perfect vision remains that of unfettered and unhampered access to the world in the clear light of the bright sun of reason. Even those who sigh that "unfortunately" we cannot be "totally freed" from the colored glasses of biases and prejudices have the same imaginary goal as those who still believe that we could indeed, by breaking away from all attachments to societies, standpoints, and feelings, access the things themselves. "If only," they all say, "we could do away with all these intermediary means with which science must abase itself in order to work—instruments, laboratories, institutions, controversies, papers, collections, theories, money [the five loops I sketched in Chapter 3]—the gaze of science would be so much more penetrating . . ." If only science could exist without what science studies relentlessly shows to be its lifeblood, how much more accurate its view of the world would be!

But this is not at all what Pasteur alludes to when he abruptly shifts from the full admission of his prejudices to the full certainty that the ferment is a living creature out there. The last thing Pasteur wants is to have his work erased and taken for a useless distortion! How could he move from a chair in Lille to a more powerful position in Paris if this were the case? On the contrary, he is extraordinarily proud of being the first in history to have artificially created the conditions to make the lactic acid ferment free to appear, at last, as a specific entity. Far from opposing filters to an unmediated gaze, it is as if *the more filters there were the clearer the gaze was,* a contradiction that the venerable optical metaphors cannot sustain without breaking down.

We might then try to shift to an industrial metaphor. When, for instance, a student of industry insists that there have been a multitude of transformations and mediations between the oil trapped deep in the geological seams of Saudi Arabia and the gas I put into the tank of my car from the old pump in the little village of Jaligny in France, the claim to reality of the gas is in no way decreased. On the contrary, it is clearly *because* of those many transformations, transportations, chemical refinements, and so on that we are able to make use of the reality of the oil, which, without all those mediations, would remain forever inaccessible to us, as safely buried as Ali Baba's treasure. There is thus a great superiority of the industrial metaphor over the optical one, of gas over gaze, to make a horrible pun: it allows one to take each intermediary step *positively*—and is well in keeping with the notion of circulating reference, a continuous circuit that should never be interrupted if the flow of information is not to break down. Either we refuse the transformations, in which case the gas remains oil far away, or we accept the transformations, but then we have gasoline and not oil!

Pasteur, however, does not have any such quasi-industrial process in mind. He does not wish to say that the lactic acid ferment is a sort of *raw material* out of which, through many clever manipulations, he has been able to refine some useful and powerful argument to convince his colleagues, and that, if the flow of connections is not interrupted, he will deliver the proof of what he says. The inadequacy of the gaze metaphor does not mean that the gas one will suffice, because it breaks down as easily as the other in the face of the bizarre nature of the phenomenon I want to highlight: the more *Pasteur* works, the more *independent* is the substance on which he works. Far from being a raw material out of which fewer and fewer features are conserved, it begins as a barely visible entity and takes on more and more competences and attributes until it ends up as a full-fledged substance! We do not simply want to say that the ferment is constructed and real as all artifacts are, but that it is *more* real *after* being transformed, as if, uncannily, there were more oil in Saudi Arabia because there is more gas in the tank of my car. Obviously the industrial metaphor of fabrication cannot handle that strange relation.

Metaphors having to do with roads, paths, or trails are slightly better because they keep the positive aspect of the intermediary transformations without touching the autonomy of the object. If we say that the laboratory experiment "paves the way" for the ferment to appear, we obviously do not imply any negation of the existence of that which is eventually reached. If we point out to the soil scientists of Chapter 2 that the cotton thread spewed out by the Topofil Chaix "leads to" their field site, they will not consider this the exposure of a "filter" that "distorts" their view, since without this little implement they would be entirely unable to follow a safe path through the Amazon forest. With the metaphor of trails, all the elements that were, so to speak, *vertical,* interposing themselves between the gaze of the researchers and their objects, become *horizontal.* What the optical metaphor forced us to take as successive veils hiding the thing, the trail metaphor lays down as so many red carpets that the researchers will walk effortlessly to access the phenomenon. We thus seem able to combine the advantage of the industrial metaphor (that all intermediaries are positive proofs of an entity's reality) with that of the gaze metaphor (that phenomena are out there, and are not the raw material for our conceptual refinery).

Alas, this is not yet the solution to Pasteur's puzzle. Despite what the metaphor of "trails" implies, phenomena are not "out there" waiting for a researcher to access them. Lactic acid ferments have to be *made visible* by Pasteur's work (just as Pasteur's philosophical innovation has to be made visible by *my* work, since this was as invisible before my intervention as the ferment was before his!). The optical metaphor may account for the visible but not for the "making" of something visible. The industrial metaphor may explain why something is "made" but not why it has thus become "visible." The trail metaphor is good at stressing the work of the scientists and their movements, but it remains as hopelessly classical as the optical one when it describes what the object is doing, that is, nothing at all, just waiting for the light to fall on it, or for the trail blazed by scientists to lead to its stubborn existence. The stage metaphor is good at pointing out that there are two planes of reference at once, but is incapable of focusing simultaneously on both, except by making the first plane the "real" backstage that allows the fiction to be believed on stage. But we do not want more fiction and more belief; we want more reality and more knowledge!

The weaknesses and benefits of all these metaphors are summarized in Figure 4.3. Each metaphor contributes to our understanding of science, but each forces us to miss important aspects of the difficulties raised by Pasteur's doubled epistemology. Pasteur points to an entirely different phenomenon that should imply at least four contradictory specifications—contradictory, that is, as long as we stick to the modernist theory of action (see Chapter 9): (1) the lactic acid ferment is wholly independent of any human construction; (2) it has no independent existence outside the work done by Pasteur; (3) this work should not be taken negatively as so many doubts about its existence, but positively as what makes it possible to exist; (4) finally, the experiment is an event and not the mere recombination of a fixed list of already present ingredients.

According to this recapitulation, experimental practice seems to be unspeakable. It does not benefit, in public parlance, from any ready-made figure of speech. The reason for this impossibility will appear

Metaphor	Benefits	Weaknesses
Parallelogram	Explains why knowledge is neither just natural nor just social	Cannot focus on the two planes at once since they are contradictory
Theater	Shows the two planes at once	Aestheticizes and shifts toward fiction even more
Fetish	Accounts for why the work is forgotten	Transforms the scientist into a dupe of his own false consciousness
Optical	Focuses attention on the independent thing	Says nothing of the work and takes all mediations as defects to be erased
Industrial	Links reality to the transformations	Takes things as raw material, losing features along the way
Trail	Turns every mediation into what makes possible the access to things	Does not modify the position of the thing sitting there and undergoing no event
Articulation	Stresses the independence of the thing; reveals the two planes at once; maintains the character of historical event; ties reality to the amount of work	Is not registered in a commonsense metaphor; leads to a set of tricky metaphysical difficulties (see Chapter 5)

Figure 4.3

later, in Chapter 7. It arises from the strange politics by which facts have been made at once completely mute and so talkative that, as the saying goes, "they speak for themselves"—thus providing the great political advantage of shutting down human babble with a voice from nowhere that renders political speech forever empty. To escape the defects of all these metaphors, we have to abandon the division between a speaking human and a mute world. As long as we have words—or gaze—on one side and a world on the other, there is no possible figure of speech that can simultaneously fulfill all four specifications; hence the misrecognition from which science studies has suffered in the public mind.

But things may be different now that, instead of the huge vertical gap between things and language, we have many small differences between horizontal paths of reference—themselves considered as series of progressive and traceable transformations, according to the lesson of Chapter 2. As is usual with science studies, common sense is no help at first and I will have to make do with my own poor resources—such as another of my inscrutable doodles. What I have been groping toward, from the beginning of this book, is an alternative to the model of statements that posits a world "out there" which language tries to reach through a correspondence across the yawning gap separating the two—as we see at the top of Figure 4.4. If my solution appears woolly, readers should remember that I am attempting to redistribute the capacity of speech between humans and nonhumans: not a task that makes for a clear exposition! They should also remember that we have abandoned, as largely illusory, the demarcation between ontological and epistemological questions, which produces much of what passes for analytical clarity.

I'd like to establish an entirely different model for the relations between humans and nonhumans by borrowing a term from Alfred North Whitehead, the notion of *propositions** (Whitehead [1929] 1978). Propositions are not statements, or things, or any sort of intermediary between the two. They are, first of all, actants*. Pasteur, the lactic acid ferment, the laboratory are all propositions. What distinguishes propositions from one another is not a *single* vertical abyss between words and the world but the *many* differences between them, without anyone knowing *in advance* if these differences are big or small, provisional or definitive, reducible or irreducible. This is precisely what the word "pro-positions" suggests. They are not positions, things, substances, or essences pertaining to a nature* made up of mute objects facing a talkative human mind, but *occasions* given to different entities to enter into contact. These occasions for interaction allow the entities to modify their definitions over the course of an event—in the present case, an experiment.

The key distinction between the two models is the role played by language. In the first model, the only way for a statement to have a reference is for it to correspond to a state of affairs. But the phrase "lactic acid ferment" does not resemble in any way the lactic acid ferment itself, any more than the word "dog" barks or the sentence "the cat is on the mat" purrs. Between the statement and the state of affairs to which

THE MODEL OF STATEMENTS

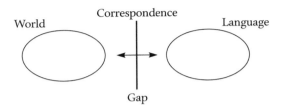

THE MODEL OF PROPOSITIONS

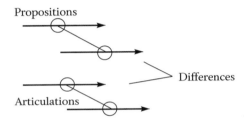

Figure 4.4 In the canonical model—see Figure 2.20—reference is obtained by bridging the gap between words and world by sending a statement across the yawning abyss and assigning it the perilous task of establishing correspondence; but if we consider neither world nor words but propositions that differ from one another, we get another relation than correspondence; the question becomes whether propositions are articulated with one another or not.

it corresponds, a radical doubt always sets in, since there should be a resemblance where none is possible. The relation established between propositions is not that of a correspondence across a yawning gap, but what I will call *articulation**. For example, Pasteur "articulates" the lactic acid ferment in his laboratory in the city of Lille. Of course this means an altogether different situation for language. Instead of being the privilege of a human mind surrounded by mute things, articulation becomes a very common property of propositions, in which many kinds of entities can participate.

Although the word is used in linguistics, articulation is in no way limited to language and may be applied not only to words but also to gestures, papers, settings, instruments, sites, trials. For instance, my friend René Boulet, in Figure 2.12, was articulating the clod of earth when inserting it into the little cardboard boxes of his "pedocomparator." If Pasteur is able to speak truthfully about the ferment, it is not because he says in words the *same thing* as what the ferment is—an impossible task since the word "ferment" does not ferment. If Pasteur, through his clever handiwork, speaks truthfully of the ferment, it is because he articulates entirely different relations for the ferment. He *proposes,* for example, that we consider it as a living and specific entity instead of as a useless by-product of a purely chemical process. In terms of what would be demanded of a corresponding statement, this is obviously a fallacy, a lie, at least a prejudice. That's exactly what Pasteur says: "I am going *beyond* that which the facts prove . . . the stand I am taking is in a framework of ideas that in rigorous terms *cannot* be irrefutably demonstrated."

Going beyond the facts and taking a stand are bad things for statements, since every trace of work and human agency obscures the goal of reaching the world out there. But they are excellent things if the aim is to articulate ever more precisely the two propositions of the lactic acid ferment and of Pasteur's laboratory. Whereas statements aim at a correspondence they can never achieve, propositions rely on the articulation of differences that make new phenomena visible in the cracks that distinguish them. Whereas statements can hope at best for sterile repetition (A is A), articulation relies on predication* with other entities (A is B, C, and so on). To say that "lactic acid fermentation," the sentence, is *like* lactic acid fermentation, the thing, does not go very far. But saying that lactic acid fermentation can be *treated like* a living organism as specific as brewer's yeast, opens up an entirely new era in the relation of science, industry, ferments, and society in the nineteenth century.

Propositions do not have the fixed boundaries of objects. They are surprising events in the histories of other entities. The more articulation there is, the better. The terms I used in the second section of this chapter, the name of actions* obtained through trials* during the event* of an experiment, now take on a different meaning. All these are ways of saying that through the artifices of the laboratory, the lactic acid ferment becomes articulable. Instead of being mute, unknown, undefined, it becomes something that is being made up of many more items, many more articles—including papers presented at the Academy!—many more reactions to many more situations. There are, quite simply, more and more things to say about it, and what is said by more and more people gains in credibility. The field of biochemistry becomes, in every sense of the term, "more articulate"—and so do the biochemists. Actually, thanks to Pasteur's ferment, they come into existence *as* biochemists, instead of having to choose between biology and chemistry as in Liebig's day. Thus we can fulfill the four specifications listed above without falling into contradiction. The more work Pasteur does, the more independent the lactic acid ferment becomes, since it is now that much more articulate, thanks to the artificial setting of the laboratory, a proposition that in no way resembles the ferment. The lactic acid ferment now exists as a discrete entity *because* it is articulated between so many others, in so many active and artificial settings.

We will flesh out this very abstract formulation in the first section of the next chapter. The point to be made now is that, in practice, it is *never* the case that we utter statements by using only the resources of language and *then* check to see if there is a corresponding thing that will verify or falsify our utterances. No one—not even the philosophers of language—has ever first said the "cat is on the mat" and *then* turned to the proverbial cat to see whether or not it is sprawled on the proverbial mat. Our involvement with the things we speak about is at once much *more intimate* and much *less direct* than that of the traditional picture: we are allowed to say new, original things when we enter well-articulated settings like good laboratories. Articulation between propositions goes much deeper than speech. We speak *because* the propositions of the world are themselves articulated, not the other way around. More exactly, *we are allowed to speak interestingly by what we allow to speak interestingly* (Despret 1996). The notion of articulated propositions establishes between knower and known entirely different relations from those in the traditional view, but it captures much more precisely the rich repertoire of scientific practice.

The Historicity of Things

Where Were Microbes before Pasteur?

"But," anyone with common sense would ask with an undertone of exasperation, "did ferments exist before Pasteur made them up?" There is no avoiding the answer: "No, they did not exist before he came along"—an answer that is obvious, natural, and even, as I will show, commonsensical! As we saw in Chapter 4, Pasteur encountered a vague, cloudy, gray substance sitting meekly in the corner of his flasks and turned it into the splendid, well-defined, articulate yeast twirling magnificently across the ballroom of the Academy. That the clock has struck twelve many times since the 1850s and her coachmen still haven't turned back into mice does nothing to change the fact that before Prince Charming came along this Cinderella was a nearly invisible by-product of a lifeless chemical process. Of course, my fairy tales aren't much more helpful than those of the science warriors who would claim that the ferment was a part of reality "out there" all along which Pasteur "discovered" with his piercing observations. No, we need not only to rethink what Pasteur and his microbes were doing before and after the experiment but to reforge the concepts that the modern settlement has given us with which to study such events. The philosophical difficulty posed by my glib response to the question above does not, however, reside in the *historicity* of ferments but in the little expression "to make up."

If we meant by "historicity" merely that our contemporary "representation" of microorganisms dates from the mid-nineteenth century, there would be no problem. We would have simply fallen back on the divide between ontological and epistemological questions that we had decided to abandon. To do away with this divide, we decided to grant historicity to the microorganisms, not only to the humans discovering them. This entails that we should be able to say that not only the microbes-for-us-humans changed in the 1850s, but also the microbes-for-themselves. Their encounter with Pasteur changed them as well. Pasteur, so to speak, "happened" to them.

If, from another perspective, we meant by "historicity" merely that the ferments "evolve over time" like the infamous cases of the flu virus or HIV, there would be no difficulty either. Like that of all living species—or for that matter, the Big Bang—the historicity of a ferment would be firmly rooted in nature. Instead of being static, phenomena would be defined as dynamic. This kind of historicity*, however, does not include the history of science and of the scientists. It is just another way to portray nature, in movement instead of as a still life. Again, the divide between what pertains to human history and what to natural history would not have been bridged in the slightest. Epistemology and ontology would remain divided, no matter how agitated or chaotic the cosmos on either side of the gap might be.

What I want to do in this chapter, halfway through this book on the reality of science studies, is to reformat the question of historicity by using the notions of proposition and articulation that I so abstractly defined at the end of the last chapter, as the only figures of speech able to fulfill all the specifications listed for Figure 4.3. What was unworkable and absurd in the subject-object fairy tale may become, if not easy, at least *thinkable* with the pair human-nonhuman. In the first section I will make an inventory of the new vocabulary we need in order to extricate ourselves from the modernist predicament—still using the same example as in Chapter 4, at the risk of giving the reader an overdose of lactic acid ferment. And then, to test the usefulness of this vocabulary, I will shift to another canonical example from Pasteur's life, his debate with Pouchet over spontaneous generation—thus descending from ferments to microbes.

Substances Have No History, but Propositions Do

I am going to submit a small series of concepts to a double torsion test, as engineers do when they verify the resistance of their materials. This will be, so to speak, my laboratory trial. We have now two lists of instruments: object, subject, gap, and correspondence on the one hand; humans, nonhumans, difference, proposition, and articulation on the other. What transformations will the notion of history undergo when put into these two different setups? What becomes feasible or unfeasible when the tension is shifted from one group of concepts to the other?

Before the notion of articulation, it was impossible to answer no to the question "Did the ferments (or the microbes) exist before Pasteur" without falling into some sort of idealism. The subject-object dichotomy distributed activity and passivity in such a way that whatever was taken by one was lost to the other. If Pasteur makes up the microbes, that is, invents them, then the microbes are passive. If the microbes "lead Pasteur in his thinking" then it is he who is the passive observer of their activity. We have begun to understand, however, that the pair human-nonhuman does not involve a tug-of-war between two opposite forces. On the contrary, the more activity there is from one, the more activity there is from the other. The more Pasteur works in his laboratory, the more autonomous his ferment becomes. Idealism was the impossible effort to give activity back to the humans, *without* dismantling the Yalta pact which had made activity a zero-sum game—and without redefining the very notion of action, as we will see in Chapter 9. In all its various forms—including of course social constructivism—idealism had a nice polemical virtue against those who granted too much independence to the empirical world. But polemics are fun to watch for only so long. If we cease to treat activity as a rare commodity of which only one team can have possession, it stops being fun to watch people trying to deprive one another of what all the players could have aplenty.

The subject-object dichotomy had another disadvantage. Not only was it a zero-sum game, but there were, by necessity, *only two* ontological species: nature and mind (or society). This rendered any account of scientific work most implausible. How could we say that in the history of ferments (Chapter 4) or that of the atomic chain reaction (Chapter 3) or that of the forest-savanna border (Chapter 2) there are only two types of actors, nature and subjects—and that, in addition, everything that one actor does not do, the second one must take over? Pasteur's culture medium, for instance: which side does it go on? Or René Boulet's pedocomparator? Or Halban's calculation of the cross-section? Do these belong to subjectivity or to objectivity or to both? None of the above, obviously, and yet each of these little mediations is indispensable for the emergence of the independent actor that is nevertheless the result of the scientists' work.

The great advantage of propositions is that they do not have to be ordered into *only two realms*. Of them it can be said without any difficulty that there are *many*. They unfold into a manifold, they don't order themselves into a duality. With the new picture I am trying to sketch, the traditional tug-of-war is dismantled twice: not only are there no winners and losers, but there are not even two teams. Thus if I say that Pasteur invents a culture medium that makes the ferment visible, I can grant activity to *all three* of the elements along the way. And if I add the Lille laboratory, then I will have *four* actors, and if I say that the Academy has been convinced, I will have *five*, and so on, without always worrying, terrified at the idea that I might run out of actors or mix up the two reserves—and the only two—from which they should be drawn.

To be sure, the subject-object dichotomy had one great superiority: it gave a clear meaning to the truth-value of a statement. A statement was said to refer if, and only if, there was a state of affairs that corresponded to it. However, as we saw in the last three chapters, this decisive advantage was turned into a nightmare when scientific practice began to be studied in detail. In spite of the thousands of books philosophers of language have thrown into the abyss separating language and world, the gap shows no sign of being filled. The mystery of reference between the two—and the only two—realms of language and world is just as obscure as before, except that we now have an incredibly sophisticated version of what happens at one pole—language, mind, brain, and now even society—and a totally impoverished version of what happens at the other, that is, *nothing*.

With propositions, one does not have to be so lopsided and sophistication may be shared equally among all the contributors to the feat of reference. Not having to fill a huge and radical gap between two realms, but merely to shift through many little gaps between slightly different active entities, reference is no longer an all-or-nothing correspondence. As we have seen often enough, the word reference* applies to the *stability* of a movement through many different implements and mediations. When we say that Pasteur speaks truthfully about a real state of affairs, we no longer ask him to jump from words to world. We say something much like the "downtown expressway moving smoothly this morning" that we hear on the radio before trying to beat the traffic. "It refers to something there" indicates the safety, fluidity, traceability, and stability of a transverse series of aligned intermediaries, not an impossible correspondence between two far-apart vertical domains. Naturally this doesn't go quite far enough, and I will have to show later how to recapture the normative distinction between truth and falsity, and at less cost, with the distinction between well-articulated and inarticulate propositions.

In any case, the sentence "The ferments existed before Pasteur made them up" means two entirely different things, depending on whether it is caught between the two poles of the subject-object dichotomy or loaded into the series of articulated humans and nonhumans. We have now reached the crux of the matter. This is where we will see if our torsion test holds up or breaks down.

In the correspondence theory of truth, the ferments are either out there or not, and if they are out there they have *always* been out there, and if they are not there they have *never* been there. They cannot appear or disappear like the flashing signals of a lighthouse. Pasteur's statements, in contrast, either correspond or do not correspond to a state of affairs and may appear or disappear according to the vagaries of history, the weight of presuppositions, or the difficulties of the task. *If we use the subject-object dichotomy, then the two—the only two—protagonists cannot share history equally.* Pasteur's statement may have a history—it appears in 1858 and not before—but the ferment cannot have such a history since it either has always been there or has never been there. Since they simply stand as the fixed target of correspondence, objects have no means of appearing and disappearing, that is, of varying.

This is the reason for the undertone of exasperation in the commonsensical question raised at the beginning of this chapter. The tension between an object with no history and statements with a history is so great that when I say "Ferments of course did not exist before 1858," I am attempting a task as impossible as holding the HMS *Britannia* at the pier with a rope after she has started steaming away. There is no sense in the expression "history of science" if we cannot somehow slacken the tension between these two poles, since we are left with only a history of scientists, while the world out there remains impervious to the other history—even if nature may still be said to be endowed with a dynamism, but that is another type of historicity altogether.

Fortunately, with the notion of circulating reference, there is nothing simpler than slackening the tension between what has and what does not have history. If the rope holding the HMS *Britannia* breaks, it is because the pier has remained fixed. But where does this fixity come from? Only from the settlement that anchors the object of reference as one extremity facing the statement on the other side across a yawning gap. "Ferments exist," however, does not qualify *one of the poles*—the pier—*but the whole series* of transformations that make up the reference. As I said, accuracy of reference indicates the fluidity and stability of a transverse series, not the bridge between two stable points or the rope between one fixed point and one that moves away. How does circulating reference help us define the historicity of things? Quite simply. *Every change* in the series of transformations that composes the reference is going to *make a difference,* and differences are all that we require, at first, to set a lively historicity into motion—as lively as a good lactic acid fermentation!

Although this sounds abstract it is much more commonsensical than the model it replaces. A lactic acid ferment grown in a culture in Pasteur's laboratory in Lille in 1858 is not the same thing as the residue of an alcoholic fermentation in Liebig's laboratory in Munich in 1852. Why not the *same* thing? Because it is not made out of the same articles, the same members, the same actors, the same implements, the same propositions. The two sentences do not repeat each other. They articulate something different. But the thing itself, where is the thing? *Here,* in the longer or shorter list of elements making it up. Pasteur is not Liebig. Lille is not Munich. The year 1852 is not the year 1858. Being sown in a culture medium is not the same as being the residue of a chemical process, and so on. The reason this answer sounds funny at first is that we still imagine the thing to be somehow at one extremity waiting out there to serve as the bedrock for the reference. But if the reference is what circulates through the whole series, every change in even *one* element of the series will make for a change in the reference. It will be a different thing to be in Lille and in Munich, to be cultivated with yeast and without it, to be seen under the microscope and with a pair of glasses, and so on.

If my slackening this tension seems like a monstrous distortion of common sense, it is because we want to have a substance* *in addition* to attributes. This is a perfectly reasonable demand since we always move from performances* to the attribution of a competence*. But as we saw in Chapter 4, the relation of substance to attributes does not have the genealogy that the subject-object dichotomy forced us to imagine: first a substance out there, outside history, and then phenomena observed by a mind. What Pasteur made clear for us—what I made clear in Pasteur's drift through multiple ontologies—is that we slowly moved from a series of attributes to a substance. The ferment began as attributes and *ended up being a substance,* a thing with clear limits, with a name, with obduracy, which was more than the sum of its parts. The word "substance*" does not designate what "remains beneath," impervious to history, but what gathers together a multiplicity of agents into a stable and coherent whole. A substance is more like the thread that holds the pearls of a necklace together than the rock bed that remains the same no matter what is built on it. In the same way that accurate reference qualifies a type of smooth and easy circulation, substance is a name that designates the *stability* of an assemblage.

This stability, however, does not have to be permanent. The best proof of this was given when, in the 1880s, to Pasteur's great surprise, enzymology took over. The ferments as living-organisms-against-Liebig's-chemical-theory again became chemical agents that could even be made through synthesis. Articulated differently they became different, and yet they were still held together by a substance, a *new* substance; they now belonged to the solid house of enzymology after having belonged for several decades, albeit in a different form, to the solid house of the emergent biochemistry.

As we shall see, the best word to designate a substance is "institution*." It made no sense to use that word before, since it clearly came from the vocabulary of social order and could not mean something other than the arbitrary imposition of a form onto matter. But in the new settlement I am outlining we are no longer prisoners of the tainted origin of such concepts. If history can be granted to ferments, substantiality can be granted to institutions. Saying that Pasteur learned through a series of routinized gestures to produce at will a lively lactic fermentation that is clearly different from the other fermentations—beer and alcohol—cannot pass for a weakening of the ferment's claim to reality. It means, on the contrary, that we are now talking about the ferment as a *matter of fact**. The state of affairs that the philosophy of language tried hopelessly to reach across the tiny bridge of correspondence resides everywhere, stolid and obdurate in the very stability of institutions. And here we have come much closer to common sense: saying that ferments began to be firmly institutionalized in Lille in 1858 surely cannot pass for anything but a truism. And saying that *they*—meaning the whole assemblage—were different in Liebig's laboratory in Munich a decade before, and that these kinds of differences are what we mean by history, certainly cannot be used as fodder for the science wars.

So we have made some progress. The negative answer to the question that opened this chapter now appears more reasonable. Associations of entities have a history if at least one of the articles making them up changes. Unfortunately, we have solved nothing yet if we do not correctly qualify the *type of historicity* that we have now distributed, with such equanimity, among all the associations making up a substance. History in itself does not guarantee that anything interesting happens. Overcoming the modernist divide is not the same thing as guaranteeing that events* will take place. If we have given a reasonable meaning to the question "Did ferments exist before Pasteur?" we are not yet through with the modernist predicament. Its sway is not only maintained by the polemical divide between subject and object, it is also enforced by the notion of causality. If history has no other meaning than to activate a potentiality*—that is, to turn into an effect what was already there, in the cause—then no matter how much juggling of associations takes place, nothing, no new thing at least, will ever happen, since the effect was *already* hidden in the cause, as a potential. Not only should science studies abstain from using society to account for nature or vice versa, it should also abstain from using causality to explain anything. Causality *follows* the events and does not precede them, as I will try to make clear in the last section of this chapter.

In the subject-object framework, ambivalence, ambiguity, uncertainty, and plasticity bothered only humans groping their way toward phenomena that were in themselves secure. But ambivalence, ambiguity, uncertainty, and plasticity also accompany creatures to which the laboratory offers the possibility of existence, a historic opportunity. If

Pasteur hesitates, we have to say that the fermentation *also* is hesitating. Objects neither hesitate nor tremble. Propositions do. Fermentation has experienced other lives before 1858 and elsewhere, but its new *concrescence**, to use another term of Whitehead's, is a unique, dated, localized life offered by Pasteur—himself transformed by his second great discovery—and by his laboratory. Nowhere in this universe— which is not of course nature*—does one find a cause, a compulsory movement, that permits one to sum up an event in order to explain its emergence. If it were otherwise, one would not be faced with an event*, with a difference, but only with the simple activation of a potential that was there all along. Time would do nothing and history would be in vain. The discovery-invention-construction of lactic yeast requires that each of the articles entering its association be given the status of a mediation*, that is, of an occurrence that is neither altogether a cause nor altogether a consequence, neither completely a means nor completely an end. As usual with philosophy, we eliminate some artificial difficulties only to encounter much trickier ones. At least these new ones are fresh and realistic—and they can be tackled empirically.

A Spatiotemporal Envelope for Propositions

If I want to render the question of where the ferments were before Pasteur commonsensical, I have to show that the vocabulary I have outlined accounts better for the history of things when they are treated just like other historical events and not as a stable bedrock above which social history unfolds and which is to be explained by appealing to already present causes. To do so I will use the debates between Louis Pasteur and Félix Archimède Pouchet over the existence of spontaneous generation. This debate is so well known that it makes a convenient site for my little experiment in comparative historiography (Farley 1972, 1974; Geison 1995; Moreau 1992; on Pouchet see Cantor 1991). The test is simple enough: are the appearance and disappearance of spontaneous generation highlighted more vividly with the dualist model or with the model of articulated propositions? Which of these two accounts fares best in our torsion test?

Let me first give a sketchy history of this case, which unfolds about four years after the one we studied in Chapter 4. Spontaneous generation was a very important phenomenon in a Europe devoid of refrigerators and ways of preserving food, a phenomenon anyone could easily reproduce in his kitchen, an undisputed phenomenon made more credible by the dissemination of the microscope. Pasteur's denial of its existence, on the contrary, existed only in the narrow confines of the laboratory on the rue d'Ulm in Paris, and only insofar as he was able to prevent, in the "swan-neck" experiment, what he called "germs carried by the air" from entering the culture flasks. When Pouchet attempted to reproduce these experiments in Rouen, the new material culture, the new skills invented by Pasteur proved too fragile to migrate from Paris to Normandy, and Pouchet found spontaneous generation occurring in his boiled flasks as readily as before.

Pouchet's difficulty in replicating Pasteur's experiments was taken as proof against Pasteur's claims, and thus as proof of the existence of the well-known universal phenomenon of spontaneous generation. Pasteur's success in *withdrawing* Pouchet's common phenomenon from space-time required a gradual and punctilious *extension* of laboratory practice to each site and each claim of his adversary. "Finally," the whole of emerging bacteriology, agro-industry, medicine, by relying on this new set of practices, eradicated spontaneous generation, transforming it into something that, although it had been a common occurrence for centuries, was now a belief in a phenomenon that "had never" existed "anywhere" in the world. This eradication, however, required the writing of textbooks, the making of historical narratives, the setting up of many institutions from universities to the Pasteur Museum, indeed an extension of each of the five loops of science's circulatory system (discussed in Chapter 3). Intense work had to be done to maintain Pouchet's claim as a belief* in a nonexistent phenomenon.

Indeed, intense work still has to be done. To this day, if you reproduce Pasteur's experiment in a defective manner by being, like me for instance, a poor experimenter, not linking your skills and material culture to the strict discipline of asepsis and germ culture learned in microbiology laboratories, the phenomena making up Pouchet's claims will still appear. Pasteurians of course will call it "contamination," and if I wrote a paper vindicating Pouchet's claims and reviving his tradition based on my observations no one would publish it. But if the collective body of precautions, the standardization, the discipline learned in Pasteurian laboratories were to be *interrupted*, not only by me, the bad experimenter, but by a whole generation of skilled technicians, then the decision about who won and who lost would become uncertain again. A society that no longer knew how to cultivate microbes and control contamination would have a hard time judging the claims of the two adversaries of 1864. There is no point in history at which a sort of inertial force can be counted on to take over the hard work of scientists and relay it into eternity. This is another extension, this time into history, of the circulating reference we began to follow in Chapter 2. For scientists there is no Day of Rest!

What interests me here is not the accuracy of this account but rather the *homology* of the narrative of the spread of microbiological skills with one that would have described, say, the rise of the Radical party from obscurity under Napoleon III to prominence in the Third Republic, or the expansion of diesel engines into submarines. The demise of Napoleon III does not mean that the Second Empire never existed, nor does the emergence of diesel engines mean that they will last forever; nor does the slow expulsion of Pouchet's spontaneous generation by Pasteur mean that it was *never* part of nature. In the same way that we may still, to this day, meet Bonapartists, although their chance of becoming President is nil, I sometimes meet spontaneous-generation buffs who defend Pouchet's claim by linking it, for instance, to prebiotics, that is, the early history of life, and who want to rewrite history yet again, although they never manage to get their "revisionist" papers published.

Both Bonapartists and spontaneous-generationists have now been pushed to the fringe, but their mere presence is an interesting indication that the "finally" that allowed philosophers of science, in the first model, to definitively rid the world of entities that had been proven wrong, was too brutal. Not only is it brutal, it also ignores the mass of work that still has to be done, daily, to activate the "definitive" version of history. After all, the Radical party disappeared, as did the Third Republic in June 1940, for lack of massive investments in democratic culture, which, like microbiology, has to be taught, practiced, kept up, made to sink in. It is always dangerous to imagine that at some point in history, *inertia* is enough to keep up the reality of phenomena that have been so difficult to produce. When a phenomenon "definitely" exists this does not mean that it exists forever, or *independently* of all practice and discipline, but that it has been entrenched in a costly and massive institution* which has to be monitored and protected with great care.

So, in the metaphysics of history that I want to substitute for the traditional one, we should be able to talk calmly about *relative existence**. It may not be the sort of existence science warriors want for objects in nature*, but it is the sort of existence science studies would like propositions to enjoy. Relative existence means that we follow the entities without stretching, framing, squeezing, and cutting them with the four adverbs never, nowhere, always, everywhere. If we use these adverbs, Pouchet's spontaneous generation will *never* have been there *anywhere* in the world; it was an illusion all along; it is not allowed to have been part of the population of entities making up space and time. Pasteur's ferments carried by the air, however, had *always* been there, all along, *everywhere,* and were bona fide members of the population of entities making up space and time long before Pasteur.

To be sure, in this kind of framework historians can tell us a few amusing things about why Pouchet and his supporters wrongly believed in the existence of spontaneous generation, and why Pasteur fumbled around for a few years before finding the right answer, but the tracing of those zigzags will give us no new essential information about the entities in question. Although they provide information on the subjectivity and history of *human* agents, history, in such a rendering, does not apply to nonhumans. By asking an entity to exist—or more exactly to have existed—either nowhere and never, or always and everywhere, the old settlement limits historicity to subjects and bans it for nonhumans. And yet existing somewhat, having a little reality, occupying a definite place and time, having predecessors and successors, these are the typical ways of delimiting what I will call the *spatiotemporal envelope** of propositions.

But why does it seem so difficult to share historicity equally among all the actors and to draw around them the envelope of relative existence without adding or subtracting anything? Because the history of science, like history proper, is embroiled in a moral issue that we have to tackle first—before we can deal later, in Chapters 7 and 8, with the even stronger political issue at stake. If we purge our accounts of the four absolute adverbs, historians, moralists, and epistemologists are afraid we may be forever unable to qualify the truth or the falsity of statements.

What do the Fafner of never-anywhere and the Fasolt of always-everywhere assert—or more exactly roar threateningly, those two giants in charge of protecting the treasure in the Nibelungen saga? That science studies has embraced a simple-minded relativism by claiming that all arguments are historical, contingent, localized, temporal, and thus cannot be differentiated, any of them being able, given enough time, to revise the others into nonexistence. Without their help, the giants boast, only an undifferentiated sea of equally valid claims will appear, engulfing at once democracy, common sense, decency, morality, and nature. The only way, according to them, to escape relativism is to *withdraw* from history and locality every fact that has been proven right, and to *stock* them safely in a nonhistorical nature* where they have always been and can no longer be reached by any sort of revision. *Demarcation** between what has and what does not have a history is, for them, the key to virtue. For this reason, historicity is granted only to humans, radical parties and emperors, while nature is periodically purged of all nonexistent phenomena. In this demarcationist view, history is simply a provisional way for humans to access nonhistorical nature: it is a convenient intermediary, a necessary evil, but it should not be, according to the two treasure guards, a *sustained mode of existence for facts*.

These claims, although they are often made, are both inaccurate and dangerous. Dangerous because, as I have said, they forget to *pay the price* of keeping up the institutions that are necessary for maintaining facts durably in existence, relying instead on the cost-free inertia of ahistoricity. But, more important, they are *inaccurate*. Nothing is easier than to differentiate in great detail the claims of Pasteur and Pouchet. This differentiation, contrary to the claims of our brawny guards, is even more telling once we abandon the boasting and empty privilege they want nonhumans to hold over human events. For science studies, *demarcation is the enemy of differentiation**. The two giants behave like the eighteenth-century French aristocrats who claimed that civil society would crash if it were not solidly supported upon their noble spines but were delegated to the humble shoulders of commoners. As it happens, civil society is better carried upon the many shoulders of the citizens than by the Atlas-like contortions of those pillars of cosmological and social order. It seems that the same demonstration can be made for differentiating the spatiotemporal envelopes deployed by science studies in redistributing activity and historicity among all of the entities involved. The common historians seem to do a much better job than the towering epistemologists of maintaining the crucial local differences.

Let us, for instance, map the two destinies of Pouchet's and Pasteur's claims, to show how clearly they can be differentiated provided they are not demarcated. Although technology as such is not an issue here—it will be in the next chapter—it may be helpful to give a rudimentary model of propositions and articulations that uses some of the tools developed to follow technological projects*. Since there is no major metaphysical difficulty in granting to diesel engines and subway systems a relative existence only, the history of technology is very much more "relaxed" than that of science as far as relative existence is concerned. Historians of technical systems know that they can have their cake (reality) and eat it too (history).

In Figure 5.1, existence is not an all-or-nothing property but a relative property which is conceived of as the *exploration* of a two-dimensional space made by association and substitution, AND and OR. An entity gains in reality if it is associated with many others that are viewed as collaborating with it. It loses in reality if, on the contrary, it has to shed associates or collaborators (human and nonhuman). Thus this figure does not include any final stage in which historicity will be surpassed, with the entity *relayed into eternity by inertia, ahistoricity, and naturalness*—although well-known phenomena like blackboxing, socialization, institutionalization, standardization, and training would be able to account for the seamless and ordinary ways in which they would be sustained and perpetuated. As we saw earlier, states of affairs become matters of fact, and then matters of course. At the bottom of Figure 5.1, the reality of Pasteur's airborne germs is obtained through an ever greater number of elements with which it is associated—machines, gestures, textbooks, institutions, taxonomies, theories, and so on. The same terms can be applied to Pouchet's claims which at version $n + 2$, time $t + 2$, are weak because they have lost almost all of their reality. The difference, so important to our two giants, between Pasteur's extended reality and Pouchet's shrinking reality can now be adequately visualized. This difference is only *as big* as the relation between the tiny segment on the left and the long segment at the right. It is *not* an *absolute* demarcation between what has never been

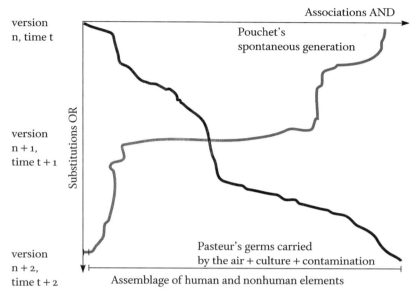

version
n, time t

Associations AND

Pouchet's
spontaneous generation

Substitutions OR

version
n + 1,
time t + 1

version
n + 2,
time t + 2

Pasteur's germs carried
by the air + culture + contamination

Assemblage of human and nonhuman elements

Figure 5.1 Relative existence may be mapped according to two dimensions: association (AND), that is, how many elements cohere together at a given time, and
substitution (OR), that is, how many elements in a given association have to be
modified to allow other new elements to cohere with the project. The result is a
curve in which every modification in the associations is "paid for" by a move in
the other dimension. Pouchet's spontaneous generation becomes less and less
real, and Pasteur's culture method becomes more and more real after undergoing
many transformations.

there and what was always there. Both are relatively real and relatively
existent, that is, extant. We never say "it exists" or "it does not exist,"
but "this is the collective history that is enveloped by the expression
spontaneous generation, or germs carried by the air."

EXHIBIT A

Let us assume that any entity is defined by an association profile
of other entities called actors. Let us suppose that those actors are
drawn from a list that ranks them, for instance, in alphabetical order.
Let us further assume that each association, called a program, is

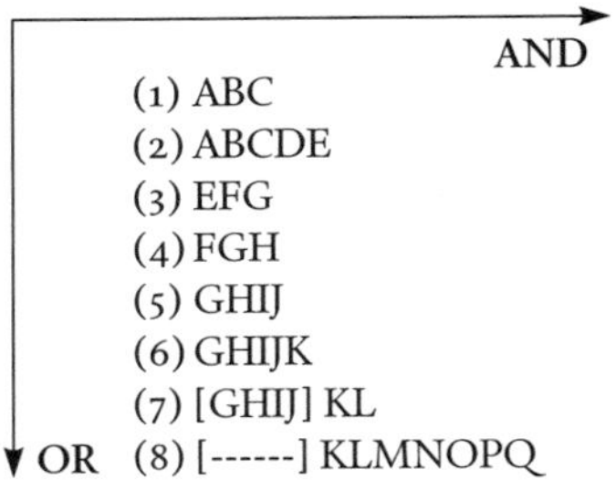

Figure A.1

counteracted by anti-programs* which dismantle or ignore the association under consideration. Finally, let us suppose that each element, in order to move from the anti-program to the program, requires some elements to leave the program and some, with which it has been already durably associated, to accompany it (Latour, Mauguin, et al. 1992).

We shall now define two intersecting dimensions: association* (akin to the linguist's syntagm*) and substitution* (or paradigm* for the linguists). To simplify, we can think of these as the AND dimension, which will be our horizontal axis, and the OR dimension, which will be our vertical axis. Any innovation can be traced both by its position on the AND–OR axes and by comparison with the record of the AND and OR positions that have successively defined it. If we replace, as a convention, all of the different actors with different letters, we can then trace the path taken by an entity, according to a progression such as the one in Figure A.1.

The vertical dimension corresponds to the exploration of substitutions, and the horizontal dimension corresponds to the number of actors that have attached themselves to the innovation (by convention we read these diagrams from top to bottom).

Each historical narrative can then be coded as follows: From X's point of view, between version (1), at time (1), and version (2), at time (2), the program ABC is transformed into ABCDE.

Then the dynamic of the narrative can be coded as follows:

To recruit F into the program, ABCD has to leave and G has to enter, which yields version (3) at time (3): EFG.

After several such versions the elements that stick together are said to "exist": they can be blackboxed together and given an identity, that is, a label, as for instance is the case for the syntagm [GHIJ], after version (7), named an institution*. The elements that have been disassociated through the different versions are said to have lost existence.

To define an entity, one will not look for an essence, or for a correspondence with a state of affairs, but for the list of all the syntagms or associations into which one element enters. This nonessentialist definition will allow for a considerable range of variations, just as a word is defined by the list of its usages: "air" will be different when associated with "Rouen" and "spontaneous generation" than when associated with "rue d'Ulm," "swan-neck experiment," and "germs"; it will mean "transport of life-force" in one case and "transport of oxygen and transport of dust-carrying germs" in the other; but the Emperor will also be different when associated by Pouchet with "ideological support of spontaneous generation to maintain God's creative power" and by Pasteur with "monetary support of laboratories without any implication about the subject matters of science." What is the essence of air? All of these associations. Who is the Emperor? All of these associations.

To make a judgment about the relative existence or nonexistence of an association, for instance "the present Emperor of France is bald," one will compare this version with others and "calculate" the stability of the association in other syntagms: "Napoleon III, Emperor of France, has a moustache," "the President of France is bald," "hairdressers have no panacea against baldness," "linguistic philosophers like to use the sentence 'the present King of France is bald.'" The length of the associations, and the stability of the connections through various substitutions and shifts in point of view, make for a great deal of what we mean by *existence* and reality.

At first sight such an opening of reality to every entity seems to defy common sense, since Golden Mountains, phlogiston, unicorns, bald kings of France, chimeras, spontaneous generation, black holes, cats on mats, and other black swans and white ravens will all occupy the same space-time as Hamlet, Popeye, and Ramses II. Such equanimity seems certainly too democratic to avoid the dangers of relativism; but this criticism forgets that our definition of existence and reality is extracted, not from a one-to-one correspondence between an isolated statement and a state of affairs, but from the unique signature drawn by associations and substitutions through the conceptual space.

As has been shown so many times by science studies, it is the *collective history* that allows us to judge the relative existence of a phenomenon; there is no higher court that would be *above* the collective and *beyond* the reach of history, although much of philosophy was devised to invent just such a court (see Chapter 7). This sketchy diagramming of narratives simply aims at directing our attention toward an alternative that does not abandon the moral aims of differentiation: each relative existence has one typical envelope and only one.

The second dimension is the one that captures historicity. History of science does not document the travel *through* time of an already existing *substance*. Such a move would accept too much of what the giants demand. Science studies documents the modifications of the ingredients that compose an articulation of entities. Pouchet's spontaneous generation, for instance, at the beginning is made of many elements: commonsense experience, anti-Darwinism, republicanism, Protestant theology, natural history, skills for observing egg development, a geological theory of multiple creations, the equipment of Rouen's natural history museum, and so on. In encountering Pasteur's opposition, Pouchet alters many of these elements. Each alteration, substitution, or translation means a movement up or down the vertical dimension of Figure 5.1. To associate elements into a durable whole, and thus to gain existence, he has to modify the list that makes up his phenomenon. But the new elements will not necessarily cohere with the earlier ones, in which case there will be a movement downward on the figure—because of the substitution—and there may be a shift to the left because of a lack of associations between the newly "recruited" elements.

For example, Pouchet has to learn a great deal of the laboratory practice of his adversary in order to fulfill the requirements of the commission nominated by the Academy of Science to adjudicate the dispute. If he fails to live up to those requirements, he loses the support of the Academy in Paris and has to rely more and more on republican scientists in the provinces. His associations may be extended—for instance he gains a great deal of support from the anti-Bonapartist popular press—but the support he expected from the Academy vanishes. The compromise between associations and substitutions is what I call *exploring the collective*. Any entity is such an exploration, such a series of events, such an experiment, such a proposition of what holds with what, of who holds with whom, of who holds with what, of what holds with whom. If Pouchet accepts the experiments of his adversary but loses the Academy and gains the popular anti-establishment press, his entity, spontaneous generation, will be a *different* entity. It is not a single substance spanning the nineteenth century unchanged. It is a set of associations, a syntagm, made of shifting compromises, a paradigm*—in the linguistic sense of the word, not the Kuhnian one—exploring what the nineteenth-century collective can withstand.

To Pouchet's dismay, there seems to be no way, working in Rouen, he can keep all his actors united in a single coherent network: Protestantism, republicanism, the Academy, boiling flasks, eggs emerging *de novo*, his ability as a natural historian, his theory of catastrophic creation. More precisely, if he wants to maintain this assemblage he has to shift audiences and give his association a completely different space and time. It now becomes a fiery battle against official science, Catholicism, bigotry, and the hegemony of chemistry over sound natural history. We should not forget that Pouchet is not doing fringe science, but is *pushed to the fringe*. At the time, it is Pouchet who seems to be able to control what is scientific by insisting that the "great problems" of spontaneous generation should be tackled only by geology and world history, not by going through Pasteur's flasks and narrow concerns.

Pasteur also explores the collective of the nineteenth century, but his association is made of elements that, at the beginning, are largely distinct from those of Pouchet. He has just started to fight Liebig's chemical theory of fermentation, as we saw in Chapter 4. This newly emerging syntagm* includes many elements: a modification of vitalism against chemistry, a reemployment of crystallographic skills such as sowing and cultivating entities, a position in Lille with many connections to agrobusiness relying on fermentation, a brand-new laboratory, experiments in making life out of inert material, a circuitous move to reach Paris and the Academy, and so on. If the ferments that Pasteur is learning to cultivate in different media, each with its own specificity—one for alcoholic fermentation, another for lactic fermentation, a third for butyric fermentation—can also be allowed to appear spontaneously, as Pouchet claims, then this will be the end of the association of the entities Pasteur has already assembled. Liebig will turn out to be right in saying that Pasteur regresses to vitalism; cultures in a pure medium will become impossible because of uncontrollable contamination; contamination itself will have to be reformatted to become the genesis of the new life forms observable under the microscope; agrobusiness will no longer be interested in a laboratory practice as haphazard as its own, and so on.

In this sketchy description I am not treating Pasteur differently from Pouchet, as if the former were struggling with real uncontaminated phenomena and the second with myths and fancies. Both try their best to hold together as many elements as they can in order to gain reality. But these are not the *same* elements. Anti-Liebig, anti-Pouchet microorganisms will authorize Pasteur to maintain the living cause of fermentation and the specificity of ferments, allowing him to control and cultivate them inside the highly disciplined and artificial limits of the laboratory, thus connecting at once with the Academy of Science and agrobusiness. Pasteur too is exploring, negotiating, trying out what holds with what, who holds with whom, what holds with whom, who holds with what. There is no other way to gain reality. But the associations he chooses and the substitutions he explores make for a different socio-natural assemblage, and each of his moves modifies the definition of the associated entities: the air as well as the Emperor, the use of laboratory equipment as well as the interpretation of preserves (that is, preserved foods), the taxonomy of microbes as well as the projects of agrobusiness.

The Institution of Substance

I have shown that we can sketch Pasteur's and Pouchet's moves in a symmetrical fashion, recovering as many differences as we wish between them without using the demarcation between fact and fiction. I have also offered a rudimentary map so as to replace judgments about existence or nonexistence with the comparison of the spatiotemporal envelopes drawn when registering associations and substitutions, syntagms and paradigms. What do we gain by this move? Why should anyone prefer science studies' account of the relative existence of all entities over the notion of a substance existing there forever? Why should adding the strange assumption of the historicity of things to the historicity of people simplify the narratives of both?

The first advantage is that we do not have to consider certain entities such as ferments, germs, or eggs sprouting into existence as being radically different from a *context* made of colleagues, emperors, money, instrument, bodily skills, and so on. The doubt about the distinction between context and content, which we disputed at the end of Chapter 3, now has the metaphysics of its ambition. Each assemblage that makes up a version in Figure A.1 is a list of heterogeneous associations that includes human and nonhuman elements. There are many philosophical difficulties with this way of arguing, but, as we saw in the case of Joliot, it has the great advantage of not requiring us to stabilize either the list of what makes up nature or the list of what makes up society. This is a decisive advantage which overcomes most of the possible defects, since, as we will see later, nature* and society* are the artifacts of a totally different political mechanism, one that has nothing to do with the accurate description of scientific practice. The less familiar the terms we use to describe human and nonhuman associations are to the subject-object dichotomy, the better.

Just as historians are not forced to imagine one single nature about which Pasteur and Pouchet would make different "interpretations," neither are they forced to imagine a single nineteenth century imposing its imprint on historical actors. What is at stake in each of the two assemblages is what God, the Emperor, matter, eggs, vats, colleagues, and so on are able to do. Each element is to be defined by its associations and is an event created at the occasion of each of those associations. This is true for the lactic acid ferment, as well as for the city of Rouen, the Emperor, the laboratory on the rue d'Ulm, God, and Pasteur and Pouchet's own standing, psychology, and presuppositions. The airborne ferments are deeply modified by the laboratory on the rue d'Ulm, but so is Pasteur, who becomes Pouchet's conqueror, and *so is the air* that is now differentiated, thanks to the eventful swan-neck experiment, into the medium that transports oxygen on the one hand, and the medium that carries dust and germs on the other.

The second advantage, as I suggested, is that we do not have to treat the two envelopes asymmetrically by considering that Pouchet is fumbling in the dark with nonexisting entities, while Pasteur is slowly homing in on an entity playing hide-and-seek, while the historians punctuate the search with warnings like "You're cold," "You're getting warmer," "Now you're hot!" We will see in Chapter 9 how this symmetry will help us bypass the impossible notion of belief. The difference between Pouchet and Pasteur is not that the first believes and the second knows. Both Pasteur and Pouchet are associating and substituting elements, very few of which are similar, and experimenting with the contradictory requirements of each entity. The associations assembled by both protagonists are similar simply in that each of them draws a spatiotemporal envelope that remains locally and temporally situated and empirically observable. The demarcation can be safely reapplied to the little differences between the entities with which Pasteur and Pouchet associate themselves, and not to the one big difference between believers and knowers.

Third, this similarity does not mean that Pasteur and Pouchet are building the *same* networks and share the *same* history. The elements in the two associations have almost no intersection—apart from the experimental setting designed by Pasteur and taken over by Pouchet until he fled in the face of the harsh demands of the Academy commission. Following the two networks in detail will lead us to completely different definitions of the nineteenth-century collective. This means that the incommensurability of the two positions—an incommensurability that seems so important for making a moral as well as an epistemological judgment—is itself the *product* of the slow differentiation of the two assemblages. Yes, in the end—a local and provisional end—Pasteur's and Pouchet's positions have been rendered incommensurable. There is no difficulty in recognizing the differences between the two networks once their basic similarity has been accepted. The spatiotemporal envelope of spontaneous generation has limits as sharp and as precise as those of germs carried by the air which contaminate microbe cultures. The abyss between the claims that our two giants compelled us to admit under threat of punishment is indeed there, but with an added bonus: *the definitive line of demarcation at which history stopped and natural ontology took over has disappeared.* As we will see in the final chapters of this book, the implementation of this line of demarcation is now ready to be analyzed for the first time independently of the problems of describing an event. In other words, we have freed differentiation from its kidnapping by a moral and political debate that had nothing to do with it.

This advantage is important because it allows us to go on qualifying, situating, and historicizing even the *extension* of a "final" reality. When we say that Pasteur has defeated Pouchet, and that now germs carried by the air are "everywhere," this everywhere can be documented em-

pirically. Viewed from the Academy of Science, spontaneous generation disappeared in 1864 through Pasteur's work. But partisans of
spontaneous generation persisted a long time and were convinced that
they had overthrown Pasteur's chemical "dictatorship"—as they called
it—and forced it to retreat into the fragile fortress of "official science."
According to them they had the field to themselves, even though Pasteur and his colleagues felt the same way. Now we can compare the
two "extended fields" without attributing the difference to that between incompatible and untranslatable "paradigms"—in the Kuhnian
sense, this time—which would forever estrange Pasteur from Pouchet.
Republicans, provincials, and natural historians who have access to
the popular anti-Bonapartist press maintain the extension of spontaneous generation. A dozen laboratories of microbiology *withdraw* the
existence of spontaneous generation from nature and reformat the
phenomena it was made of by the twin practices of pure medium cultivation and protection against contamination. The two are not incompatible paradigms. They were *made* incompatible by the series
of associations and substitutions of each of the two assemblies of protagonists. They simply began to have fewer and fewer elements in
common.

Why we may find this reasoning difficult is that we imagine microbes must have a substance that is a little bit *more* than the series of
its historical manifestations. We may be ready to grant that the set of
performances always remains inside the networks and that they are
delineated by a precise spatiotemporal envelope, but we cannot sup-
press the feeling that the substance travels with fewer constraints than
the performances. It seems to live a life of its own, having been, like
the Virgin Mary in the dogma of Immaculate Conception, always already there, even before Eve's fall, waiting in Heaven to be implanted
in Anne's womb at the right time. There is indeed a *supplement* in the
notion of substance, but it is better accounted for, as I suggested in the
first section of this chapter, by the notion of institution*.

Such a reworking of the notion of substance is crucial because it
points to something that is badly accounted for by the history of science: how do phenomena *remain in existence* without a law of inertia?
Why can't we say that Pasteur was right and Pouchet was wrong?
Well, we can say it, but only on the condition that we render very
clearly and precisely the institutional mechanisms that are *still at work*

to maintain the asymmetry between the two positions. The solution to this problem is to formulate the question in the following way: In whose world are we now living, that of Pasteur or that of Pouchet? I don't know about you, but for my part, I live inside the Pasteurian network, every time I eat pasteurized yogurt, drink pasteurized milk, or swallow antibiotics. In other words, to account for even a long-lasting victory, one does not have to grant extrahistoricity to a research program as if it would suddenly, at some threshold or turning point, need *no* further upkeep. What was an event must remain a continuing event. One simply has to go on historicizing and localizing the network and finding who and what make up its descendants.

In this sense I participate in the "final" victory of Pasteur over Pouchet, in the same way that I participate in the "final" victory of republican over autocratic modes of government by voting in the next presidential election instead of abstaining or refusing to register. To claim that such a victory requires no further work, no further action, no further institution, would be foolish. I can simply say that I have inherited Pasteur's microbes, I am the descendant of this event, which in turn depends on what I make of it today (Stengers 1993). To claim that the "everywhere and always" of such events covers the whole spatiotemporal manifold would be at best an exaggeration. Step away from the present networks, and completely different definitions of yogurt, milk, and forms of government will be generated, and this time, not spontaneously . . . The scandal is not that science studies preaches relativism but that, in the science wars, those who claim that the labor of keeping up the institutions of truth can be interrupted *without risk* pass for paragons of morality. We will understand later how they accomplished this little trick and managed to turn the tables of morality on us.

The Puzzle of Backward Causation

There are still, I am well aware, many loose ends in this generalized use of the notions of event and proposition to replace expressions like "discovery," "invention," "fabrication," or "construction." One of them is the very notion of construction borrowed from technical practice, which we are going to deconstruct, so to speak, in the next chapter. Another one is the glib answer that I gave at the beginning of the chapter to the question "Did microbes exist before Pasteur?" I claimed that my answer, "Of course not," was commonsensical. I cannot end this chapter without demonstrating why I think it is so.

What does it mean to say that there were microbes "before" Pasteur? Contrary to the first impression, there is no deep metaphysical mystery in this long time "before" Pasteur, but only a very simple optical illusion that disappears as soon as the work of extending existence *in time* is documented as empirically as its extension *in space*. My solution, in other words, is to historicize more, not less. No sooner had Pasteur stabilized his theory of germs carried by the air than he reinterpreted the practices of the past in a new light, saying that what went wrong in the fermentation of beer, for example, was the inadvertent contamination of the vats by other ferments:

> Whenever an albuminous liquid of a suitable nature contains a substance such as sugar, capable of undergoing diverse chemical transformations dependent upon the nature of such and such a ferment, the germs of these ferments all *tend* to propagate at the same time, and usually they develop simultaneously, unless one of the ferments *invades* the medium more rapidly than the others. *It is precisely this last circumstance that one determines when one uses this method of sowing an organism* that is already formed and ready to reproduce. (§16)

It is now possible, for Pasteur, to understand retrospectively what farming and industry have been doing all along without knowing it. The difference between past and present is that Pasteur now masters the culture of organisms instead of unwittingly being manipulated by invisible phenomena. Sowing germs in a culture medium is the rearticulation by Pasteur of what others before him, not understanding what it was, named disease, invasion, or mishap. The art of lactic acid fermentation becomes a laboratory science. In the laboratory, conditions may be mastered at will. In other words, Pasteur *reinterpreted* the past practices of fermentation as fumbling around in the dark with entities against which one could now protect oneself.

How has this retrospective vision of the past been achieved? What Pasteur did was to produce in 1864 a new version of the years 1863, 1862, 1861, which now included a new element: "microbes fought unwittingly by faulty and haphazard practices." Such a retroproduction of history is a familiar feature for historians, especially historians of history (Novick 1988). There is nothing easier to understand than how Christians, after the first century, reformatted the entire Old Testament as confirmation of a long and hidden preparation for Christ's birth, or how European nations had to reinterpret the history of German culture after the end of World War II. Exactly the same thing happened with Pasteur. He *retrofitted* the past with his own microbiology: the year 1864 that was built *after* 1864 did not have the same components, textures, and associations as the year 1864 produced *during 1864*. I try to make this point as simple as possible in Figure 5.2.

If this enormous work of retrofitting—which includes history telling, textbook writing, instrument making, body training, and the creation of professional loyalties and genealogies—is ignored, then the question "Did the microbes exist before Pasteur?" takes on a paralyzing aspect that can stupefy the mind for a minute or two. After a few minutes, however, the question becomes empirically answerable: Pasteur also took care to *extend* his local production into other times and places and to make the microbes the *substrate* of other people's unwitting actions. We now understand better the curious etymology of the word "substance," which has been causing us so much trouble in these two chapters on Pasteur. Substance does not mean that there is a durable and ahistorical "substrate" *behind* the attributes, but that it is possible, because of the sedimentation of time, to turn a new entity into what *lies beneath other entities*. Yes, there are substances that have been there all along, but on the condition that they are made the substrate of activities, in the past as well as in space. So there are two practical meanings now given to the word substance*: one is the institution* that holds together a vast array of practical setups, as we saw earlier, and the other is the work of *retrofitting* that situates a more recent event as what "lies beneath" an older one.

The "everywhere and always" may be reached, but it is costly, and its localized and temporal extension remains visible throughout. It may take a while before we can effortlessly juggle all these dates (and dates of dates), but there is no logical inconsistency in talking about the extension in time of scientific networks, any more than there are discrepancies in following their extension in space. It can even be said that the difficulties in handling these apparent paradoxes are tiny compared with the smallest of those offered by relativistic physics. If science had not been kidnapped for entirely different ends, there

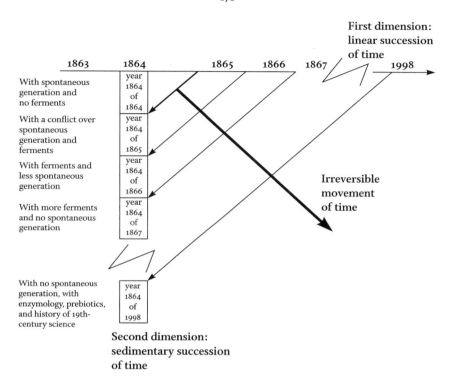

Figure 5.2 Time's arrow is the resultant of two dimensions, not one: the first dimension, the linear succession of time, always moves forward (1865 is *after* 1864); the second one, sedimentary succession, moves backward (1865 occurs *before* 1864). When we ask the question "Where was the ferment before 1865?" we do not reach the top segment of the column that makes up the year 1864, but only the transverse line that marks the contribution of the year 1865 to the elaboration of the year 1864. This, however, implies no idealism or backward causation, since time's arrow always moves irreversibly forward.

would be no difficulty in describing the appearance and disappearance of propositions that never stopped having a history. Now that we have begun to see that scientific practice can be studied, we are equipped to find the motives behind this kidnapping and even the culprits' hideout. But before we can do this we still have one long detour to make, by way of the master of detours: Daedalus the engineer. Without beginning to rework part of the philosophy of technology and part of the myth of progress, we won't be able to shake off the moral and political burden that the modernist settlement has so unfairly placed on the shoulders of nonhumans. Nonhumans are born free, and everywhere they are in chains.

EXHIBIT B

A year should be defined along two axes, not only one. The first axis registers the linear dimension of time, that is, the succession of years. In that sense 1864 happens *before* 1865. But this is not all there is to say about the year 1864. A year is not only a figure in a series of integers, it is also a column along a second axis that registers the sedimentary succession of time. In this second dimension there is also a portion of what happened in 1864 that is produced *after* 1864 and made retrospectively a part of the ensemble that forms, from then on, the sum of what happened in the year 1864.

In the case represented in Figure 5.2, the year 1864 is formed of as many segments as there have been years since. If the year 1864 "of 1864" contains spontaneous generation as a generally accepted phenomenon, the year 1864 "of 1865" includes, in addition, an intense conflict over spontaneous generation. This conflict no longer rages another year later, after the scientific community has definitively accepted Pasteur's theory of airborne germs. The year 1864 "of 1866" thus includes a vestigial belief in spontaneous generation and a triumphant Pasteur.

This process of sedimentation never ends. If we skip forward 130 years, there is still a year 1864 "of 1998," to which has been added many features, not only a rich new historiography of the dispute between Pasteur and Pouchet, but maybe also a complete revision of the dispute in which, eventually, Pouchet is the winner because he anticipated some results of prebiotics.

What gives an appearance of depth to the question "Where were the airborne germs before 1864?" is a very simple confusion between the first, linear dimension of time and the second, sedimentary dimension. If one considers only the first dimension the answer is "nowhere," since the first segment in the column that makes up the whole year 1864 does *not* include any airborne germs. The consequence is not, however, an absurd form of idealism, since most of the other sedimentary segments of the year 1864 *do* include airborne germs. It is thus possible to say, without contradiction, both "Airborne germs were made up in 1864" and "They were there all along," that is, *all along* the vertical column that recapitulates all the components of the year 1864 produced since.

In that sense, no more fundamental objections are raised by the question "Where were microbes before Pasteur?" than by this other question, which nobody would even think of raising: "Where was Pasteur before 1822 (the year of his birth)?"

I am thus arguing that the only commonsensical answer to the question is "After 1864 airborne germs were there all along." This solution involves treating extension in time as rigorously as extension in space. To be everywhere in space or always in time, work has to be done, connections made, retrofitting accepted.

If the answers to these apparent puzzles are so straightforward, then the question is no longer whether to take such "mysteries" seriously, but why people take them as deep philosophical puzzles that would condemn science studies to absurdity.

A Collective of Humans and Nonhumans

Following Daedalus's Labyrinth

The Greeks used to distinguish the straight path of reason and scientific knowledge, *episteme*, from the clever and crooked path of technical know-how, *metis*. Now that we have seen how indirect, devious, mediated, interconnected, vascularized are the paths taken by scientific facts, we may be able to find a different genealogy for technical artifacts as well. This is all the more necessary because so much of science studies relies on the notion of "construction," borrowed from technical action. As we are going to see, however, the philosophy of technology is no more directly useful for defining human and nonhuman connections than epistemology has been, and for the same reason: in the modernist settlement, theory fails to capture practice, for a reason that will only become clear in Chapter 9. Technical action, thus, presents us with puzzles as bizarre as those involved in the articulation of facts. Having grasped how the classical theory of objectivity fails to do any justice to the practice of science, we are now going to see that the notion of "technical efficiency over matter" in no way accounts for the subtlety of engineers. We may then be able, finally, to understand these nonhumans, which are, I have been claiming since the beginning, full-fledged actors in our collective; we may understand at last why we do not live in a society gazing out at a natural world or in a natural world that includes society as one of its components. Now that nonhumans are no longer confused with objects, it may be possible to imagine the collective in which humans are entangled with them.

In the myth of Daedalus, all things deviate from the straight line. After Daedalus's escape from the labyrinth, Minos used a subterfuge worthy of Daedalus himself to find the clever craftsman's hiding place and take revenge. Minos, in disguise, heralded far and wide his offer of a reward to anyone who could thread the circumvoluted shell of a snail. Daedalus, hidden at the court of King Cocalus and unaware that the offer was a trap, managed the trick by replicating Ariadne's cunning: he attached a thread to an ant and, after allowing it to enter the shell through a hole at its apex, he induced the ant to weave its way through this tiny labyrinth. Triumphant, Daedalus claimed his reward, but King Minos, equally triumphant, asked for Daedalus's extradition to Crete. Cocalus abandoned Daedalus; still, this artful dodger managed, with the help of Cocalus's daughters, to divert the hot water from the plumbing system he had installed in the palace, so that it fell, as if by accident, on Minos in his bath. (The king died, boiled like an egg.) Only for a brief while could Minos outwit his master engineer—Daedalus was always one ruse, one machination ahead of his rivals.

Daedalus embodies the sort of intelligence for which Odysseus (of whom the *Iliad* says that he is *polymetis,* a bag of tricks) is most famed (Détienne and Vernant 1974). Once we enter the realm of engineers and craftsmen, no unmediated action is possible. A *daedalion,* the word in Greek that has been used to describe the labyrinth, is something curved, veering from the straight line, artful but fake, beautiful but contrived (Frontisi-Ducroux 1975). Daedalus is an inventor of contraptions: statues that seem to be alive, military robots that watch over Crete, an ancient version of genetic engineering that enables Poseidon's bull to impregnate Pasiphae to conceive the Minotaur—for which he builds the labyrinth, from which, via another set of machines, he manages to escape, losing his son Icarus on the way. Despised, indispensable, criminal, ever at war with the three kings who draw their power from his machinations, Daedalus is the best eponym for technique—and the concept of *daedalion* is the best tool for penetrating the evolution of what I have called so far the collective*, which in this chapter I want to define more precisely. Our path will lead us not only through philosophy but through what could be called a *pragmatogony** that is, a wholly mythical "genesis of things," in the fashion of the cosmogonies of the past.

Folding Humans and Nonhumans into Each Other

To understand techniques—technical means—and their place in the collective, we have to be as devious as the ant to which Daedalus attached his thread (or as the worms bringing the forest to the savanna in Chapter 2). The straight lines of philosophy are of no use when it is the crooked labyrinth of machinery and machinations, of artifacts and *daedalia,* that we have to explore. To cut a hole at the apex of the shell and weave my thread, I need to define, in opposition to Heidegger, what mediation means in the realm of techniques. For Heidegger a technology is never an instrument, a mere tool. Does that mean that technologies mediate action? No, because we have ourselves become instruments for no other end than instrumentality itself (Heidegger 1977). Man—there is no Woman in Heidegger—is possessed by technology, and it is a complete illusion to believe that we can master it. We are, on the contrary, framed by this *Gestell,* which is one way in which Being is unveiled. Is technology inferior to science and pure knowledge? No, because, for Heidegger, far from serving as applied science, technology dominates all, even the purely theoretical sciences. By rationalizing and stockpiling nature, science plays into the hands of technology, whose sole end is to rationalize and stockpile nature without end. Our modern destiny—technology—appears to Heidegger radically different from *poesis,* the kind of "making" that ancient craftsmen knew how to achieve. Technology is unique, insuperable, omnipresent, superior, a monster born in our midst which has already devoured its unwitting midwives. But Heidegger is mistaken. I will try to show why by using a simple, well-known example to demonstrate the impossibility of speaking of any sort of mastery in our relations with nonhumans, *including* their supposed mastery over us.

"Guns kill people" is a slogan of those who try to control the unrestricted sale of guns. To which the National Rifle Association replies with another slogan, "Guns don't kill people; *people* kill people." The first slogan is materialist: the gun acts by virtue of *material* components irreducible to the social qualities of the gunman. On account of the gun the law-abiding citizen, a good guy, becomes dangerous. The

NRA, meanwhile, offers (amusingly enough, given its political views) a *sociological* version more often associated with the Left: that the gun does nothing in itself or by virtue of its material components. The gun is a tool, a medium, a neutral carrier of human will. If the gunman is a good guy, the gun will be used wisely and will kill only when appropriate. If the gunman is a crook or a lunatic, then, with *no change in the gun itself*, a killing that would in any case occur will be (simply) carried out more efficiently. What does the gun add to the shooting? In the materialist account, *everything:* an innocent citizen becomes a criminal by virtue of the gun in her hand. The gun enables, of course, but also instructs, directs, even pulls the trigger—and who, with a knife in her hand, has not wanted at some time to stab someone or something? Each artifact has its script, its potential to take hold of passersby and force them to play roles in its story. By contrast, the sociological version of the NRA renders the gun a *neutral* carrier of will that *adds nothing* to the action, playing the role of a passive conductor, through which good and evil are equally able to flow.

I have caricatured the two positions, of course, in an absurdly diametrical opposition. No materialist would really claim that guns kill by themselves. What the materialist claims, more exactly, is that the good citizen is *transformed* by carrying the gun. A good citizen who, without a gun, might simply be angry may become a criminal if he gets his hands on a gun—as if the gun had the power to change Dr. Jekyll into Mr. Hyde. Materialists thus make the intriguing suggestion that our qualities as subjects, our competences, our personalities, depend on what we hold in our hands. Reversing the dogma of moralism, the materialists insist that we are what we have—what we have in our hands, at least.

As for the NRA, its members cannot truly maintain that the gun is so neutral an object that it has no part in the act of killing. They have to acknowledge that the gun *adds* something, though not to the moral state of the person holding it. For the NRA, one's moral state is a Platonic essence: one is born either a good citizen or a criminal. Period. As such, the NRA account is moralist—what matters is what you are, not what you have. The sole contribution of the gun is to speed the act. Killing by fists or knives is simply slower, dirtier, messier. With a gun, one kills better, but at no point does the gun modify one's goal. Thus NRA sociologists make the troubling suggestion that we can master techniques, that techniques are nothing more than pliable and diligent slaves. This simple example is enough to show that artifacts are no easier to grasp than facts: it took us two chapters to understand Pasteur's doubled epistemology, and it is going to take us a long time to understand precisely what things make us do.

The First Meaning of Technical Mediation: Interference

Who or what is responsible for the act of killing? Is the gun no more than a piece of mediating technology? The answer to these questions depends on what mediation* means. A first sense of mediation (I will offer four) is what I will call the *program of action**, the series of goals and steps and intentions that an agent can describe in a story like the one about the gun and the gunman (see Figure 6.1). If the agent is human, is angry, wants to take revenge, and if the accomplishment of the agent's goal is interrupted for whatever reason (perhaps the agent is not strong enough), then the agent makes a *detour*, a deviation like the one we saw in Chapter 3 in the operations of conviction between Joliot and Dautry: one cannot speak of techniques any more than of science without speaking of *daedalia*. (Although in English the word "technology" tends to replace the word "technique," I will make use of both terms throughout, reserving the tainted term "technoscience" for a very specific stage in my mythical pragmatogony.) Agent 1 falls back on Agent 2, here a gun. Agent 1 enlists the gun or is enlisted by it—it does not matter which—and a third agent emerges from a fusion of the other two.

The question now becomes which goal the new composite agent will pursue. If it returns, after its detour, to Goal 1, then the NRA story obtains. The gun is then a tool, merely an intermediary. If Agent 3 drifts from Goal 1 to Goal 2, then the materialist story obtains. The gun's intent, the gun's will, the gun's script have superseded those of Agent 1; it is human action that is no more than an intermediary. Note that in the figure it makes no difference if Agent 1 and Agent 2 are reversed. The myth of the Neutral Tool under complete human control and the myth of the Autonomous Destiny that no human can master are symmetrical. But a third possibility is more commonly realized: the creation of a new goal that corresponds to neither agent's program of action. (You only wanted to injure but, with a gun now in your

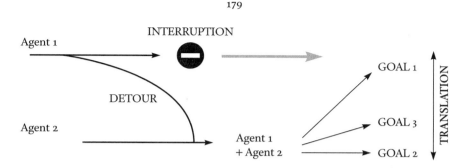

FIRST MEANING OF MEDIATION : GOAL TRANSLATION

Figure 6.1 As in Figure 3.1, we can portray the relation between two agents as a translation of their goals which results in a composite goal that is different from the two original goals.

hand, you want to kill.) In Chapter 3 I called this uncertainty about goals translation*. As should be clear by now, translation does not mean a shift from one vocabulary to another, from one French word to one English word, for instance, as if the two languages existed independently. I used translation to mean displacement, drift, invention, mediation, the creation of a link that did not exist before and that to some degree modifies the original two.

Which of them, then, the gun or the citizen, is the *actor* in this situation? *Someone else* (a citizen-gun, a gun-citizen). If we try to comprehend techniques while assuming that the psychological capacity of humans is forever fixed, we will not succeed in understanding how techniques are created nor even how they are used. You are a different person with the gun in your hand. As Pasteur showed us in Chapter 4, essence is existence and existence is action. If I define you by what you have (the gun), and by the series of associations that you enter into when you use what you have (when you fire the gun), then you are modified by the gun—more so or less so, depending on the weight of the other associations that you carry.

This translation is wholly symmetrical. You are different with a gun in your hand; the gun is different with you holding it. You are another subject because you hold the gun; the gun is another object because it has entered into a relationship with you. The gun is no longer the gun-in-the-armory or the gun-in-the-drawer or the gun-in-the pocket, but the gun-in-your-hand, aimed at someone who is screaming. What is true of the subject, of the gunman, is as true of the object, of the gun that is held. A good citizen becomes a criminal, a bad guy becomes a worse guy; a silent gun becomes a fired gun, a new gun becomes a used gun, a sporting gun becomes a weapon. The twin mistake of the materialists and the sociologists is to start with essences, those of subjects or those of objects. As we saw in Chapter 5, that starting point renders impossible our measurement of the mediating role of techniques as well as those of science. If we study the gun and the citizen as propositions, however, we realize that neither subject nor object (nor their goals) is fixed. When the propositions are articulated, they join into a new proposition. They become "someone, something" else.

It is now possible to shift our attention to this "someone else," the hybrid actor comprising (for instance) gun and gunman. We must learn to attribute—redistribute—actions to many more agents than are acceptable in either the materialist or the sociological account. Agents can be human or (like the gun) nonhuman, and each can have goals (or functions, as engineers prefer to say). Since the word "agent" in the case of nonhumans is uncommon, a better term, as we have seen, is actant*. Why is this nuance important? Because, for example, in my vignette of the gun and the gunman, I could replace the gunman with "a class of unemployed loiterers," translating the individual agent into a collective; or I could talk of "unconscious motives," translating it into a subindividual agent. I could redescribe the gun as "what the gun lobby puts in the hands of unsuspecting children," translating it from an object into an institution or a commercial network; or I could call it "the action of a trigger on a cartridge through the intermediary of a spring and a firing-pin," translating it into a mechanical series of causes and consequences. These examples of actor-actant symmetry force us to abandon the subject-object dichotomy, a distinction that prevents the understanding of collectives. It is neither people nor guns that kill. Responsibility for action must be shared among the various actants. And this is the first of the four meanings of mediation.

The Second Meaning of Technical Mediation: Composition

One might object that a basic asymmetry lingers—women make computer chips, but no computer has ever made women. Common sense, however, is not the safest guide here, any more than it is in the sciences. The difficulty we just encountered with the example of the gun remains, and the solution is the same: the prime mover of an action becomes a new, distributed, and nested series of practices whose sum may be possible to add up but only if we respect the mediating role of all the actants mobilized in the series.

To be convincing on this point will require a short inquiry into the way we talk about tools. When someone tells a story about the invention, fabrication, or use of a tool, whether in the animal kingdom or the human, whether in the psychological laboratory or the historical or the prehistoric, the structure is the same (Beck 1980). Some agent has a goal or goals; suddenly the access to the goal is interrupted by that breach in the straight path that distinguishes *metis* from *episteme*. The detour, a *daedalion*, begins (Figure 6.2). The agent, frustrated, turns around in a mad and random search, and then, whether by insight or eureka or by trial and error (there are various psychologies available to account for this moment) the agent seizes upon some other agent—a stick, a partner, an electrical current—and then, so the story goes, returns to the previous task, removes the obstacle, and achieves the goal. Of course, in most tool stories there is not one but two or several *subprograms** nested in one another. A chimpanzee

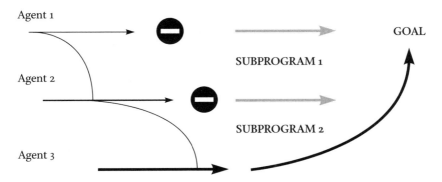

SECOND MEANING OF MEDIATION : COMPOSITION

Figure 6.2 If the number of subprograms is increased, then the composite goal— here the thick curved line—becomes the common achievement of each of the agents bent by the process of successive translation.

might seize a stick and, finding it too blunt, begin, after another crisis, another subprogram, to sharpen the stick, inventing en route a compound tool. (How far the multiplication of these subprograms can continue raises interesting questions in cognitive psychology and evolutionary theory.) Although one can imagine many other outcomes—for instance, the loss of the original goal in the maze of subprograms)—let us suppose that the original task has been resumed.

What interests me here is the *composition* of action marked by the lines that get longer at each step in Figure 6.2. Who performs the action? Agent 1 plus Agent 2 plus Agent 3. Action is a property of associated entities. Agent 1 is allowed, authorized, enabled, afforded by the others. The chimp plus the sharp stick reach (not reaches) the banana. The attribution to one actor of the role of prime mover in no way weakens the necessity of a composition of forces to explain the action. It is by mistake, or unfairness, that our headlines read "Man flies," "Woman goes into space." Flying is a property of the whole association of entities that includes airports and planes, launch pads and ticket counters. B-52s do not fly, the U.S. Air Force flies. Action is simply not a property of humans *but of an association of actants,* and this is the second meaning of technical mediation. Provisional "actorial" roles may be attributed to actants only because actants are in the process of exchanging competences, offering one another new possibilities, new goals, new functions. Thus symmetry holds in the case of fabrication as it does in the case of use.

But what does symmetry mean? Symmetry is defined by what is conserved through transformations. In the symmetry between humans and nonhumans, I keep constant the series of competences, of properties, that agents are able to swap by overlapping with one another. I want to situate myself at the stage *before* we can clearly delineate subjects and objects, goals and functions, form and matter, before the swapping of properties and competences is observable and interpretable. Full-fledged human subjects and respectable objects out there in the world cannot be my starting point; they may be my point of arrival. Not only does this correspond to the notion of articulation* I explored in Chapter 5, but it is also consistent with many well-established myths that tell us that we have been made by our tools. The expression *Homo faber* or, better, *Homo faber fabricatus* describes, for Hegel and André Leroi-Gourhan (Leroi-Gourhan 1993) and Marx and Bergson, a dialectical movement that ends by making us sons and daughters of our own works. As for Heidegger, the relevant myth is that "So long as we represent technology as an instrument, we remain held fast in the will to master it. We press on past the essence of technology" (Heidegger 1977, p. 32). We will see later what can be done with dialectics and the *Gestell*, but if inventing myths is the only way to get on with the job, I shall not hesitate to make up a new one and even to throw in a few more of my diagrams.

The Third Meaning of Technical Mediation: The Folding of Time and Space

Why is it so difficult to measure, with any precision, the mediating role of techniques? Because the action that we are trying to measure is subject to blackboxing*, a process that makes the joint production of actors and artifacts entirely opaque. Daedalus's maze shrouds itself in secrecy. Can we open the labyrinth and count what is inside?

Take, for instance, an overhead projector. It is a point in a sequence of action (in a lecture, say), a silent and mute intermediary*, taken for granted, completely determined by its function. Now suppose the projector breaks down. The crisis reminds us of the projector's existence. As the repairmen swarm around it, adjusting this lens, tightening that bulb, we remember that the projector is made of several parts, each with its role and function and its relatively independent goals. Whereas a moment before the projector scarcely existed, now even its parts have individual existence, each its own "black box." In an instant our "projector" grew from being composed of zero parts to one to many. How many actants are really there? The philosophy of technology we need has little use for arithmetic.

The crisis continues. The repairmen fall into a routinized sequence of actions, replacing parts. It becomes clear that their actions are composed of steps in a sequence that integrates several human gestures. We no longer focus on an object but see a group of people gathered *around* an object. A shift has occurred between actant and mediator.

Figures 6.1 and 6.2 showed that goals are redefined by associations with nonhuman actants, and that action is a property of the whole association, not only of those actants called human. However, as Figure 6.3 shows, the situation is even more confused, since the *number* of

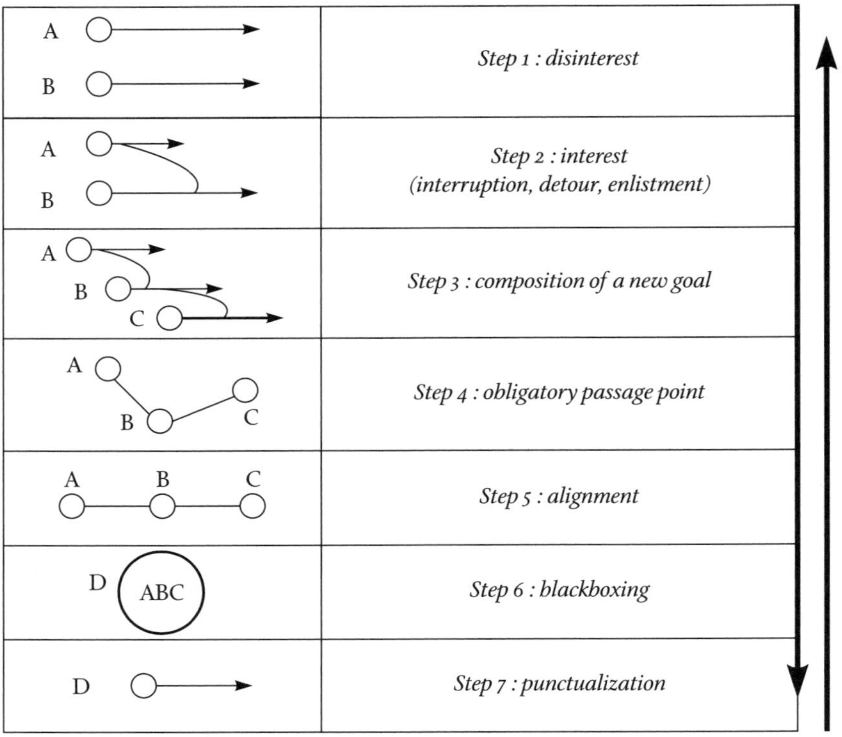

A ⟶ B ⟶	*Step 1 : disinterest*
A B ⟶	*Step 2 : interest* *(interruption, detour, enlistment)*
A B C ⟶	*Step 3 : composition of a new goal*
A B C	*Step 4 : obligatory passage point*
A B C ○—○—○	*Step 5 : alignment*
D (ABC)	*Step 6 : blackboxing*
D ○⟶	*Step 7 : punctualization*

THIRD MEANING OF MEDIATION:
REVERSIBLE BLACKBOXING

Figure 6.3 Any given assembly of artifacts may be moved up or down this succession of steps depending on the crisis they go through. What we may consider, in routine use, as one agent (step 7) may turn out to be composed of several (step 6) that may not even be aligned (step 4). The history of the earlier translations they had to go through may become visible, until they are freed again from any influence of the others (step 1).

actants varies from step to step. The composition of objects also varies: sometimes objects appear stable, sometimes they appear agitated, like a group of humans around a malfunctioning artifact. Thus the projector may count for one part, for nothing, for one hundred parts, for so many humans, for no humans—and each part itself may count for one, for zero, for many, for an object, for a group. In the seven steps of Figure 6.3, each action may proceed toward either the dispersion of actants or their integration into a single punctuated whole (a whole that, soon thereafter, will count for nothing). We need to account for all seven steps.

Look around the room in which you are puzzling over Figure 6.3. Consider how many black boxes there are in the room. Open the black boxes; examine the assemblies inside. Each of the parts inside the black box is itself a black box full of parts. If any part were to break, how many humans would immediately materialize around each? How far *back* in time, *away* in space, should we retrace our steps to follow all those silent entities that contribute peacefully to your reading this chapter at your desk? Return each of these entities to step 1; imagine the time when each was disinterested and going its own way, without being bent, enrolled, enlisted, mobilized, folded in any of the others' plots. From which forest should we take our wood? In which quarry should we let the stones quietly rest?

Most of these entities now sit in silence, as if they did not exist, invisible, transparent, mute, bringing to the present scene their force and their action from who knows how many millions of years past. They have a peculiar ontological status, but does this mean that they do not act, that they do not mediate action? Can we say that because we have made all of them—and who is this "we," by the way? not I, certainly—should they be considered slaves or tools or merely evidence of a *Gestell?* The depth of our ignorance about techniques is unfathomable. We are not even able to count their number, nor can we tell whether they exist as objects or as assemblies or as so many sequences of skilled actions. Yet there remain philosophers who believe there are such things as abject objects . . . If science studies once believed that relying on the construction of artifacts would help account for facts, it is in for a surprise. Nonhumans escape the strictures of objectivity twice; they are neither objects known by a subject nor objects manipulated by a master (nor, of course, are they masters themselves).

The Fourth Meaning of Technical Mediation: Crossing the Boundary between Signs and Things

The reason for such ignorance is made clearer when we consider the fourth and most important meaning of mediation. Up to this point I have used the terms "story" and "program of action," "goal" and "function," "translation" and "interest," "human" and "nonhuman," as if techniques were dependable denizens that support the world of discourse. But techniques modify the matter of our expression, not only its form. Techniques have meaning, but they produce meaning via a special type of articulation that, once again, like the circulating reference we met in Chapter 2 and the variable ontology we followed in Chapter 4, crosses the commonsense boundary between signs and things.

Here is a simple example of what I have in mind: the speed bump that forces drivers to slow down on campus, which in French is called a "sleeping policeman." The driver's goal is translated, by means of the speed bump, from "slow down so as not to endanger students" into "slow down and protect your car's suspension." The two goals are far apart, and we recognize here the same displacement as in our gun story. The driver's first version appeals to morality, enlightened disinterest, and reflection, whereas the second appeals to pure selfishness and reflex action. In my experience, there are many more people who would respond to the second than to the first: selfishness is a trait more widely distributed than respect for law and life—at least in France! The driver modifies his behavior through the mediation of the speed bump: he falls back from morality to force. But from an observer's point of view it does not matter through which channel a given behavior is attained. From her window the chancellor sees that cars are slowing down, respecting her injunction, and for her that is enough.

The transition from reckless to disciplined drivers has been effected through yet another detour. Instead of signs and warnings, the campus engineers have used concrete and pavement. In this context the notion of detour, of translation, should be modified to absorb, not only (as with previous examples) a shift in the definition of goals and functions, but also *a change in the very matter of expression*. The engineers' program of action, "make drivers slow down on campus," is now articulated with concrete. What would the right word be to account for this articulation? I could have said "objectified" or "reified" or "realized" or "materialized" or "engraved," but these words imply an all-powerful human agent imposing his will on shapeless matter, while nonhumans also act, displace goals, and contribute to their definition.

As we see, it is no easier to find the right term for the activity of techniques than for the efficacy of the lactic acid ferments—we will understand in Chapter 9 that this is because they are all factishes*. In the meantime I want to propose yet another term, *delegation* (see Figure 6.4).

Not only has one meaning, in the example of the speed bump, been displaced into another, but an action (the enforcement of the speed law) has been translated into another kind of expression. The engineers' program is delegated in concrete, and in considering this shift we leave the relative comfort of linguistic metaphors and enter unknown territory. We have not abandoned meaningful human relations and abruptly entered a world of brute material relations—although this might be the impression of drivers, used to dealing with negotiable signs but now confronted by nonnegotiable speed bumps. The shift is not from discourse to matter because, for the engineers, the speed bump is one *meaningful articulation* within a gamut of propositions from which they are no more free to choose than the syntagms* and paradigms* we saw in Chapter 5. What they can do is to explore the associations and the substitutions that trace a unique trajectory through the collective. Thus *we remain in meaning but no longer in discourse;* yet we do not reside among mere objects. Where are we?

Before we can even begin to elaborate a philosophy of techniques we have to understand delegation as yet another type of shifting*, in

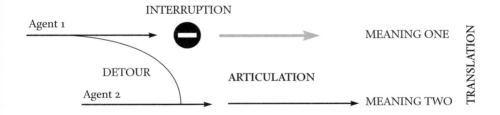

FOURTH MEANING OF MEDIATION : DELEGATION

Figure 6.4 As in Figure 6.1, the introduction of a second agent in the path of a first one implies a process of translation; but here the shift in meaning is much greater, since the very nature of the "meaning" has been modified. The matter of the expression has changed along the way.

addition to the one that we used in Chapter 4 to understand Pasteur's laboratory work. If I say to you, for instance, "Let us imagine ourselves in the shoes of the campus engineers when they decided to install the speed bumps," I not only transport you into another space and time but translate you into another actor (Eco 1979). I shift you out of the scene you now occupy. The point of spatial, temporal, and "actorial" shifting, which is basic to all fiction, is to make the reader travel without moving (Greimas and Courtés 1982). You make a detour through the engineers' office, but without leaving your seat. You lend me, for a time, a character who, with the aid of your patience and imagination, travels with me to another place, becomes another actor, then returns to become yourself in your own world again. This mechanism is called identification, by means of which the "enunciator" (I) and the "enunciatee" (you) both invest in the shifting delegates of ourselves within other composite frames of reference.

In the case of the speed bump the shift is "actorial": the "sleeping policeman," as the bump is known, is not a policeman, does not resemble one in the least. The shift is also spatial: on the campus road there now resides a new actant that slows down cars (or damages them). Finally, the shift is temporal: the bump is there night and day. But the enunciator of this technical act has disappeared from the scene—where are the engineers? where is the policeman?—while someone, something, reliably acts as lieu-tenant, holding the enunciator's place. Supposedly the co-presence of enunciators and enunciatees is necessary for an act of fiction to be possible, but what we now have is an absent engineer, a constantly present speed bump, and an enunciatee who has become the user of an artifact.

One may object that this comparison between fictional shifting and the shifts of delegation in technical activity is spurious: to be transported in imagination from France to Brazil is not the same as taking a plane from France to Brazil. True enough, but where does the difference reside? With imaginative transportation, you simultaneously occupy all frames of reference, shifting into and out of all the delegated *personae* that the storyteller offers. Through fiction, *ego, hic, nunc* may be shifted, may become other *personae,* in other places, at other times. But aboard the plane I cannot occupy more than one frame of reference at a time (unless, of course, I sit back and read a novel which takes me, say, to Dublin on a fine June day in 1904). I am seated in an object-institution that connects two airports through an airline. The act of transportation has been *shifted down**, not out—down to planes, engines, and automatic pilots, object-institutions to which has been delegated the task of moving while the engineers and managers are absent (or limited to monitoring). The co-presence of enunciators and enunciatees has collapsed, along with their many frames of reference, to a single point in time and space. All the frames of reference of the engineers, air-traffic controllers, and ticket agents have been brought together into the single frame of reference of Air France flight 1107 to São Paulo.

An object *stands in* for an actor and creates an asymmetry between absent makers and occasional users. Without this detour, this shifting down, we would not understand how an enunciator could be absent: either it is there, we would say, or it does not exist. But through shifting down another combination of absence and presence becomes possible. In delegation it is not, as in fiction, that I am here and elsewhere, that I am myself and someone else, but that an action, long past, of an actor, long disappeared, is still active here, today, on me. I live in the midst of technical *delegates;* I am folded into nonhumans.

The whole philosophy of techniques has been preoccupied by this detour. Think of technology as *congealed* labor. Consider the very notion of investment: a regular course of action is suspended, a detour is initiated via several types of actants, and the return is a fresh hybrid that carries past acts into the present and permits its many investors to disappear while also remaining present. Such detours subvert the order of time and space—in a minute I may mobilize forces set into motion hundreds or millions of years ago in faraway places. The relative shapes of actants and their ontological status may be completely reshuffled—techniques act as *shape-changers,* making a cop out of a barrel of wet concrete, lending a policeman the permanence and obstinacy of stone. The relative ordering of presence and absence is redistributed—we hourly encounter hundreds, even thousands, of absent makers who are remote in time and space yet simultaneously active and present. And through such detours, finally, the political order is subverted, since I rely on many delegated actions that themselves make me do things on behalf of others who are no longer here, the course of whose existence I cannot even retrace.

A detour of this kind is not easy to understand, and the difficulty is compounded by the accusation of fetishism* made by critics of technology, as we will see in Chapter 9. It is us, the human makers (so they say), that you see in those machines, those implements, us under another guise, our own hard work. We should restore the human labor (so they command) that stands behind those idols. We heard this story told, to different effect, by the NRA: guns do not act on their own, only humans do so. A fine story, but it comes centuries too late. Humans are no longer *by themselves*. Our delegation of action to other actants that now share our human existence has developed so far that a program of antifetishism could only lead us to a nonhuman world, a lost, phantasmagoric world *before* the mediation of artifacts. The erasure of delegation by the critical antifetishists would render the shifting *down* to technical artifacts as opaque as the shifting *out* to scientific facts (see Figure 6.4).

But we cannot fall back on materialism either. In artifacts and technologies we do not find the efficiency and stubbornness of matter, imprinting chains of cause and effect onto malleable humans. The speed bump is ultimately *not* made of matter; it is full of engineers and chancellors and lawmakers, commingling their wills and their story lines with those of gravel, concrete, paint, and standard calculations. The mediation, the technical translation, that I am trying to understand resides in the blind spot in which society and matter exchange properties. The story I am telling is not a *Homo faber* story, in which the courageous innovator breaks away from the constraints of social order to make contact with hard and inhuman but—at last—objective matter. I am struggling to approach the zone where some, though not all, of the characteristics of pavement become policemen, and some, though not all, of the characteristics of policemen become speed bumps. I have earlier called this zone articulation*, and this is not, as I hope is now clear, a sort of golden mean or dialectic between objectivity and subjectivity. What I want to find is another Ariadne's thread—another Topofil Chaix—to follow how Daedalus folds, weaves, plots, contrives, finds solutions where none are visible, using any expedient at hand, in the cracks and gaps of ordinary routines, swapping properties among inert, animal, symbolic, concrete, and human materials.

Technical Is a Good Adjective, Technique a Lousy Noun

We now understand that techniques do not exist as such, that there is nothing that we can define philosophically or sociologically as an object, as an artifact or a piece of technology. There does not exist, any more in technology than in science, anything to play the role of the foil for the human soul in the modernist scenography. The noun "technique"—or its upgraded version, "technology"—does not need to be used to separate humans from the multifarious assemblies with which they combine. But there is an *adjective,* technical, that we can use in many different situations, and rightly so.

"Technical" applies, first of all, to a subprogram, or a series of nested subprograms, like the ones discussed earlier. When we say "this is a technical point," it means that we have to *deviate* for a moment from the main task and that we will eventually *resume* our normal course of action, which is the only focus worth our attention. A black box opens momentarily, and will soon be closed again, becoming completely invisible in the main sequence of action.

Second, "technical" designates the *subordinate* role of people, skills, or objects that occupy this secondary function of being present, indispensable, but invisible. It thus indicates a specialized and highly circumscribed task, clearly subordinate in a hierarchy.

Third, the adjective designates a hitch, a snag, a catch, a hiccup in the smooth functioning of the subprograms, as when we say that "there is a technical problem to solve first." Here the deviation may not lead us back to the main road, as with the first meaning, but may *threaten* the original goal entirely. Technical is no longer a mere detour, but an obstacle, a roadblock, the beginning of a detour, of a long translation, maybe of a whole new labyrinth. What should have been a means may become an end, at least for a while, or maybe a maze, in which we are lost forever.

The fourth meaning carries the same uncertainty about what is an end and what is a means. "Technical skill" and "technical personnel" apply to those with a unique ability, a knack, a gift, and also to the ability to make themselves *indispensable,* to occupy privileged though inferior positions which might be called, borrowing a military term, obligatory passage points. So technical people, objects, or skills are at once inferior (since the main task will eventually be resumed), indispensable (since the goal is unreachable without them), and, in a way, capricious, mysterious, uncertain (since they depend on some highly specialized and sketchily circumscribed knack). Daedalus the perverse and Vulcan the limping god are good illustrations of this meaning of technical. So the adjective technical has a useful meaning that agrees in common parlance with the first three types of mediation defined above, interference, composition of goals, and blackboxing.

"Technical" also designates a very specific type of *delegation*, of movement, of shifting down, that crosses over with entities that have a different timing, different spaces, different properties, different ontologies, and that are made to share the same destiny, thus creating a new actant. Here the noun form is often used as well as the adjective, as when we say "a technique of communication," "a technique for boiling eggs." In this case the noun does not designate a thing, but a *modus operandi*, a chain of gestures and know-how bringing about some anticipated result.

If one ever comes face to face with a technical object, this is never the beginning but the *end* of a long process of proliferating mediators, a process in which all relevant subprograms, nested one into another, meet in a "simple" task. Instead of the legendary kingdom in which subjects meet objects, one generally finds oneself in the realm of the *personne morale*, of what is called the "body corporate" or the "artificial person." Three extraordinary terms! As if the personality became moral by becoming collective, or collective by becoming artificial, or plural by doubling the Saxon word body with a Latin synonym, *corpus*. A *body corporate* is what we and our artifacts have become. We are an object-institution.

The point sounds trivial if applied asymmetrically. "Of course," one might say, "a piece of technology must be seized and activated by a human subject, a purposeful agent." But the point I am making is symmetrical: what is true of the "object" is still truer of the "subject." There is no sense in which humans may be said to exist as humans without entering into commerce with what authorizes and enables them to exist (that is, to act). A forsaken gun is a mere piece of matter, but what would an abandoned gunner be? A human, yes (a gun is only one artifact among many), but not a soldier—and certainly not one of the NRA's law-abiding Americans. Purposeful action and intentionality may not be properties of objects, but they are not properties of humans either. They are the properties of institutions, of apparatuses, of what Foucault called *dispositifs*. Only corporate bodies are able to absorb the proliferation of mediators, to regulate their expression, to redistribute skills, to force boxes to blacken and close. Objects that exist simply as objects, detached from a collective life, are unknown, buried in the ground. Technical artifacts are as far from the status of efficiency as scientific facts are from the noble pedestal of objectivity. Real artifacts are always parts of institutions, trembling in their mixed status as mediators, mobilizing faraway lands and people, ready to become people or things, not knowing if they are composed of one or of many, of a black box counting for one or of a labyrinth concealing multitudes (MacKenzie 1990). Boeing 747s do not fly, airlines fly.

Pragmatogony: Is There an Alternative to the Myth of Progress?

In the modernist settlement, objects were housed within nature and subjects within society. We have now replaced objects and subjects with scientific facts and technical artifacts, which have an entirely different destiny and shape. Whereas objects could only face out at the subjects—and vice versa—nonhumans may be folded into humans through the key processes of translation, articulation, delegation, shifting out and down. What name can we give to the house in which they have taken up residence? Not nature*, of course, since its existence is entirely polemical, as we will see in the next chapter. Society* will not do either, since it has been turned, by the social scientists, into a fairy tale of social relations, from which all nonhumans have been carefully enucleated (see Chapter 3). In the newly emerging paradigm, we have substituted the notion of collective*—defined as an exchange of human and nonhuman properties inside a corporate body—for the tainted word "society."

We Live in Collectives, Not in Societies

In abandoning dualism our intent is not to throw everything into the same pot, to efface the distinct features of the various parts within the collective. We want analytical clarity, too, but following different lines than the one drawn for the polemical tug of war between objects and subjects. The name of the game is not to extend subjectivity to things, to treat humans like objects, to take machines for social actors, but *to avoid using* the subject-object distinction *at all* in order to talk about the folding of humans and nonhumans. What the new picture seeks to capture are the moves by which any given collective *extends* its social fabric to *other* entities. This is what I have meant, until now, by the provisional expression "Science and technology are what *socialize* nonhumans to bear upon human relations." This is the makeshift expression I had forged as a substitute for the modernist expression: "Science and technology allow minds to break away from society to reach objective nature, and to impose order on efficient matter."

What I'd like is one more diagram, in which we could trace, not how human subjects can break away from the shackles of social life to impose order on nature or to retrieve natural laws to maintain order in society, but how a collective of one given definition can modify its makeup by articulating different associations. In this impossible diagram I would need to follow a series of coherent moves: first, there would be translation*, the means by which we articulate different sorts of matter; next, what I will call, borrowing an image from genetics, crossover, which consists of the exchange of properties among humans and nonhumans; third, a step that can be called enrollment, by which a nonhuman is seduced, manipulated, or induced into the collective; fourth, as we saw in the case of Joliot and his military clients, the mobilization of nonhumans inside the collective, which adds fresh unexpected resources, resulting in strange new hybrids; and, finally, displacement, the direction the collective takes once its shape, extent, and composition have been altered by the enrollment and mobilization of new actants. If we had such a diagram, we would do away with social constructivism for good. Alas, I and my Macintosh have not been able to do better than Figure 6.5.

The only advantage of this figure is to provide a basis for the comparison of collectives, a comparison that is completely independent of demography (of their scale, so to speak). What science studies has done over the past fifteen years is subverted the distinction between ancient techniques (the *poesis* of artisans) and modern (broad-scale, inhuman, domineering) technologies. This distinction was never more than a prejudice. You can modify the size of the half-circle in Figure 6.5, but you do not have to modify its shape. You can modify the angle of the tangents, the extent of the translation, the types of enrollment, the size of the mobilization, the impact of the displacement, but you

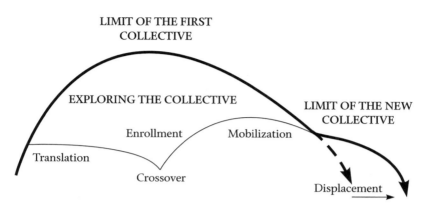

Figure 6.5 Instead of portraying science and technology as breaking away from the strict limits of a society, a collective is conceived as constantly modifying its limit through a process of exploration.

don't have to oppose those collectives that deal only with social relations and those that have been able to break away from them in order to deal with the laws of nature. Contrary to what makes Heideggerians weep, there is an extraordinary *continuity*, which historians and philosophers of technology have increasingly made legible, between nuclear plants, missile-guidance systems, computer-chip design, or subway automation and the ancient mixture of society, symbols, and matter that ethnographers and archaeologists have studied for generations in the cultures of New Guinea, Old England, or sixteenth-century Burgundy (Descola and Palsson 1996). Unlike what is held by the traditional distinction, the difference between an ancient or "primitive" collective and a modern or "advanced" one is *not* that the former manifests a rich mixture of social and technical culture while the latter exhibits a technology devoid of ties with the social order.

The difference, rather, is that the latter translates, crosses over, enrolls, and mobilizes more elements which are more intimately connected, with a more finely woven social fabric, than the former does. The relation between the scale of collectives and the number of nonhumans enlisted in their midst is crucial. One finds, of course, longer chains of action in "modern" collectives, a greater number of nonhumans (machines, automatons, devices) associated with one another, but one must not overlook the *size* of markets, the *number* of people in their orbits, the *amplitude* of the mobilization: more objects, yes, but many more subjects as well. Those who have tried to distinguish these two sorts of collective by attributing "objectivity"and "efficiency" to modern technology and "humanity" to low-tech *poesis* have been deeply mistaken. Objects and subjects are made simultaneously, and an increased number of subjects is directly related to the number of objects stirred—brewed—into the collective. The adjective modern* does not describe an *increased distance* between society and technology or their alienation, but a deepened *intimacy*, a more intricate mesh, between the two.

Ethnographers describe the complex relations implied by every technical act in traditional cultures, the long and mediated access to matter that these relations suppose, the intricate pattern of myths and rites necessary to produce the simplest adze or the simplest pot, revealing that a variety of social graces and religious mores were necessary for humans to interact with nonhumans (Lemonnier 1993). But do we, even today, have unmediated access to naked matter? Is our interaction with nature short on rites, myths, and protocols (Descola and Palsson 1996)? Has the vascularization of science diminished or increased? Has the maze of Daedalus become straighter or more convoluted?

To believe that we have modernized ourselves would be to ignore most of the cases examined by science and technology studies. How mediated, complicated, cautious, mannered, even baroque is the access to matter of any piece of technology! How many sciences—the functional equivalent of rites—are necessary to prepare artifacts for socialization! How many persons, crafts, and institutions must be in place for the enrollment of even one nonhuman, as we saw with the lactic acid ferment of Chapter 4, or the chain reaction of Chapter 3, or the soil samples of Chapter 2! When ethnographers describe our biotechnology, artificial intelligence, microchips, steelmaking, and so on, the fraternity of ancient and modern collectives is instantly obvious. If anything, what we took as merely symbolic in the old collectives is taken *literally* in the new: in contexts where a few dozen people were once required, thousands are now mobilized; where shortcuts were once possible, much longer chains of action are now necessary. Not fewer but more, and more intricate, customs and protocols, not fewer mediations but more: many more.

The most important consequence of getting beyond the *Homo faber* myth is that, when we exchange properties with nonhumans through technical delegation, we enter into a complex transaction that pertains to "modern" as well as to traditional collectives. If anything, the modern collective is the one in which the relations of humans and nonhumans are so intimate, the transactions so many, the mediations so convoluted, that there is no plausible sense in which artifact, corporate body, and subject can be distinguished. In order to take account of this symmetry between humans and nonhumans, on the one hand, and this continuity between traditional and modern collectives, on the other, social theory must be somewhat modified.

It is a commonplace in critical theory to say that techniques are social because they have been "socially constructed"—yes, I know, I also used that term once, but that was twenty years ago and I recanted it immediately, since I meant something entirely different from what sociologists and their adversaries mean by social. The notion of a social mediation is vacuous if the meanings of "mediation" and "social" are not made precise. To say that social relations are "reified" in technology, such that when we are confronted with an artifact we are confronted, in effect, with social relations, is to assert a tautology, and a very implausible one at that. If artifacts are nothing but social relations, then why must society work through them to inscribe itself in something else? Why not inscribe itself *directly*, since the artifacts count for nothing? Because, critical theorists continue, through the medium of artifacts, domination and exclusion hide themselves under the guise of natural and objective forces. Critical theory thus deploys a tautology—social relations are nothing but social relations—to which it adds a conspiracy theory: society is hiding behind the fetish of techniques.

But techniques are not fetishes*, they are unpredictable, not means but mediators, means and ends at the same time; and that is why they bear upon the social fabric. Critical theory is unable to explain why artifacts enter the stream of our relations, why we so incessantly recruit and socialize nonhumans. It is not to mirror, congeal, crystallize, or hide social relations, but to remake these very relations through fresh and unexpected sources of action. Society is not stable enough to inscribe itself in anything. On the contrary, most of the features of what we mean by social order—scale, asymmetry, durability, power, hierarchy, the distribution of roles—are impossible even to define without recruiting socialized nonhumans. Yes, *society is constructed, but not socially constructed*. Humans, for millions of years, have extended their social relations to other actants with which, with whom, they have swapped many properties, and with which, with whom, they form collectives.

A "Servant" Narrative: The Mythical History of Collectives

A detailed case study of sociotechnical networks ought to follow at this juncture, but many such studies have already been written, and most have failed to make their new social theory felt, as the science wars have made painfully clear to all. Despite the heroic efforts of these studies, many of their authors are all too often misunderstood by readers as cataloguing examples of the "social construction" of technology. Readers account for the evidence mustered in them according to the dualist paradigm that the studies themselves frequently undermine. The obstinate devotion to "social construction" as an explanatory device, whether by careless readers or "critical" authors, seems to derive from the difficulty of disentangling the various meanings of the catchword *sociotechnical*. What I want to do, then, is to peel away, one by one, these layers of meaning and attempt a genealogy of their associations.

Moreover, having disputed the dualist paradigm for years, I have come to realize that no one is prepared to abandon an arbitrary but useful dichotomy, such as that between society and technology, if it is not replaced by categories that have at least a semblance of providing the same discriminating power as the one jettisoned. Of course, I will never be able to do the same political job with the pair human-nonhuman as the subject-object dichotomy has accomplished, since it was in fact to free science from politics that I embarked on this strange undertaking, as I will make clear in the next chapters. In the meantime we can toss around the phrase "sociotechnical assemblages" forever without moving beyond the dualist paradigm that we wish to leave behind. To move forward I must convince the reader that, pending the resolution of the political kidnapping of science, *there is an alternative to the myth of progress.* At the heart of the science wars lies the powerful accusation that those who undermine the objectivity of science and the efficiency of technology are trying to lead us backward into some primitive, barbaric dark age—that, incredibly, the insights of science studies are somehow "reactionary."

In spite of its long and complex history, the myth of progress is based on a very rudimentary mechanism (Figure 6.6). What gives the thrust to the arrow of time is that modernity at last breaks out of a confusion, made in the past, between what objects really are in themselves and what the subjectivity of humans believes them to be, projecting onto them passions, biases, and prejudices. What could be called a front of modernization—like the Western Frontier—thus clearly distinguishes the confused past from the future, which will be more and more radiant, no doubt about that, because it will distinguish even more clearly the efficiency and objectivity of the laws of nature from the values, rights, ethical requirements, subjectivity, and politics of the human realm. With this map in their hands, science warriors have no difficulty situating science studies: "Since they are always insisting that objectivity and subjectivity [the science warriors' terms for nonhumans and humans] are mixed up, science students are leading us in only one possible direction, into the obscure past out of which we must extract ourselves by a movement of radical conversion,

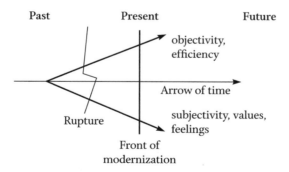

Figure 6.6 What makes the arrow of time thrust forward in the modernist narrative of progress is the certainty that the past will differ from the future because what was confused will become distinct: objectivity and subjectivity will no longer be mixed up. The result of this certainty is a front of modernization that allows one to distinguish slips backward from steps forward.

the conversion through which a barbarian premodernity becomes a civilised modernity."

In an interesting case of cartographic incommensurability, however, science studies uses an entirely different map (Figure 6.7). The arrow of time is *still there*, it still has a powerful and maybe irresistible thrust, but an entirely different mechanism makes it tick. Instead of clarifying even further the relations between objectivity and subjectivity, time enmeshes, at an ever greater level of intimacy and on an ever greater scale, humans and nonhumans with each other. The feeling of time, the definition of where it leads, of what we should do, of what war we should wage, is entirely different in the two maps, since in the one I use, Figure 6.7, the confusion of humans and nonhumans is not only our past *but our future as well*. If there is one thing of which we may be as certain as we are of death and taxation, it is that we will live tomorrow in imbroglios of science, techniques, and society *even more tightly linked* than those of yesterday—as the mad cow affair has demonstrated so clearly to European beefeaters. The difference between the two maps is total, because what the modernist science warriors see as a horror to be avoided at all costs—the mixing up of objectivity and subjectivity—is for us, on the contrary, the hallmark of a civilized life, except that what time mixes up in the future even more than in the past *are not objects and subjects at all, but humans and nonhumans,* and that makes a world of difference. Of this difference the science warriors remain blissfully ignorant, convinced that we want to confuse objectivity and subjectivity.

I am now in the usual quandary of this book. I have to offer an alternative picture of the world that can rely on none of the resources of common sense although, in the end, I aim at nothing but common sense. The myth of progress has centuries of institutionalization behind it, and my little pragmatogony is helped by nothing but my miserable diagrams. And yet I have to go on, since the myth of progress is so powerful that it puts any discussion to an end.

Yes, I want to tell another tale. For my present pragmatogony*, I have isolated eleven distinct layers. Of course I do not claim for these definitions, or for their sequence, any plausibility. I simply want to show that the tyranny of the dichotomy between objects and subjects is not inevitable, since it is possible to envision another myth in which it plays no role. If I succeed in opening some space for the imagination,

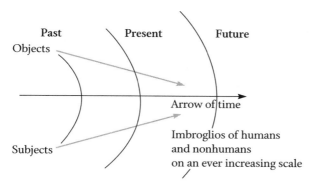

Figure 6.7 In the alternative "servant" narrative there is still an arrow of time, but it is registered very differently from Figure 6.6: the two lines of objects and subjects become more confused in the future than they were in the past, hence the feeling of instability. What is growing instead is the ever expanding scale at which humans and nonhumans are connected together.

then we are not forever stuck with the implausible myth of progress. If I could even begin to recite this pragmatogony—I use this word to insist on its fanciful character—I would have found an alternative to the myth of progress, that most powerful of all the modernist myths, the one that held my friend under its sway when he asked me, in Chapter 1, "Do we know more than we used to?" No, we don't know more, if by this expression we mean that every day we extract ourselves further from a confusion between facts, on the one hand, and society, on the other. But yes, we do know a good deal more, if by this we mean that our collectives are tying themselves ever more deeply, more intimately, into imbroglios of humans and nonhumans. Until we have an alternative to the notion of progress, provisional as it may be, science warriors will always be able to attach to science studies the infamous stigma of being "reactionary."

I will build this alternative with the strangest of means. I want to highlight the successive crossovers through which humans and nonhumans have exchanged their properties. Each of those crossovers results in a dramatic change in the scale of the collective, in its composition, and in the degree to which humans and nonhumans are enmeshed. To tell my tale I will open Pandora's box backward; that is, starting with the most recent types of folding, I will try to map the labyrinth until we find the earliest (mythical) folding. As we will see, contrary to the science warriors' fear, no dangerous regression is involved here, since all of the earlier steps are still with us today. Far from being a horrifying miscegenation between objects and subjects, they are simply the very hybridizations that make us humans and nonhumans.

Level 11: Political Ecology

Talk of a crossover between techniques and politics does not, in my pragmatogony, indicate belief in the distinction between a material realm and a social one. I am simply unpacking the eleventh layer of what is packed in the definitions of society and technique. The eleventh interpretation of the crossover—the swapping of properties—between humans and nonhumans is the simplest to define because it is the most *literal.* Lawyers, activists, ecologists, businessmen, political philosophers, are now seriously talking, in the context of our ecological crisis, of granting to nonhumans some sort of rights and even legal standing. Not so many years ago, contemplating the sky meant thinking of matter, or of nature. These days we look up at a sociopolitical imbroglio, since the depletion of the ozone layer brings together a scientific controversy, a political dispute between North and South, and immense strategic changes in industry. Political representation of nonhumans seems not only plausible now but necessary, when the notion would have seemed ludicrous or indecent not long ago. We used to deride primitive peoples who imagined that a disorder in society, a pollution, could threaten the natural order. We no longer laugh so heartily, as we abstain from using aerosols for fear the sky may fall on our heads. Like the "primitives," we fear the pollution caused by our negligence—which means of course that neither "they" nor "we" have ever been primitive.

As with all crossovers, all exchanges, this one mixes elements from both sides, the political with the scientific and technical, and this mixture is not a haphazard rearrangement. Technologies have taught us how to manage vast assemblies of nonhumans; our newest sociotechnical hybrid brings what we have learned to bear on the political system. The new hybrid remains a nonhuman, but not only has it lost its material and objective character, it has acquired properties of citizenship. It has, for instance, the right not to be enslaved. This first layer of meaning—the last in chronological sequence to arrive—is that of political ecology or, to use Michel Serres's term, "the natural contract"(Serres 1995). *Literally*, not symbolically as before, we have to manage the planet we inhabit, and must now define what I will call in the next chapter a politics of things.

Level 10: Technoscience

If I descend to the tenth layer, I see that our current definition of technology is itself due to the crossover between a previous definition of society and a particular version of what a nonhuman can be. To illustrate: some time ago, at the Institut Pasteur, a scientist introduced himself, "Hi, I am the coordinator of yeast chromosome 11." The hybrid whose hand I shook was, all at once, a person (he called himself "I"), a corporate body ("the coordinator"), and a natural phenomenon (the genome, the DNA sequence, of yeast). The dualist paradigm will not allow us to understand this hybrid. Place its social aspect on one side and yeast DNA on the other, and you will bungle not only the speaker's words but also the opportunity to grasp how a genome becomes known to an organization and how an organization is naturalized in a DNA sequence on a hard disk.

We again encounter a crossover here, but it is of a different sort and goes in a different direction, although it could also be called sociotechnical. For the scientist I interviewed there is no question of granting any sort of rights, of citizenship, to yeast. For him yeast is a strictly material entity. Still, the industrial laboratory where he works is a place in which new modes of organization of labor elicit completely new features in nonhumans. Yeast has been put to work for millennia, of course, for instance in the old brewing industry, but now it works for a network of thirty European laboratories where its genome is mapped, humanized, and socialized, as a code, a book, a program of action, compatible with our ways of coding, counting, and reading, retaining none of its material quality, the quality of an outsider. It is absorbed into the collective. Through technoscience—defined, for my purposes here, as a fusion of science, organization, and industry—the forms of coordination learned through "networks of power" (see Level 9) are extended to inarticulate entities. Nonhumans are endowed with speech, however primitive, with intelligence, foresight, self-control, and discipline, in a fashion both large-scale and intimate. Socialness is shared with nonhumans in an almost promiscuous way. While in this model, the tenth meaning of sociotechnical (see Figure 6.8), automata have no rights, they are much more than material entities; they are complex organizations.

Level 9: Networks of Power

Technoscientific organizations, however, are not purely social, because they themselves recapitulate, in my story, nine prior crossovers of humans and nonhumans. Alfred Chandler and Thomas Hughes have each traced the interpenetration of technical and social factors in what Chandler terms the "global corporation" (Chandler 1977) and Hughes terms "networks of power" (Hughes 1983). Here again the phrase "sociotechnical imbroglio" would be apt, and one could replace the dualist paradigm with the "seamless web" of technical and social factors so beautifully traced by Hughes. But the point of my little genealogy is also to identify, inside the seamless web, properties borrowed from the social world in order to socialize nonhumans and properties borrowed from nonhumans in order to naturalize and expand the social realm. For each layer of meaning, whatever happens happens as if we are learning, in our contacts with one side, ontological properties that are then reimported to the other side, generating new, completely unexpected effects.

The extension of networks of power in the electrical industry, in telecommunications, in transportation, is impossible to imagine without a massive mobilization of material entities. Hughes's book is exemplary for students of technology because it shows how a technical invention (electric lighting) led to the establishment (by Edison) of a corporation of unprecedented scale, its scope directly related to the physical properties of electrical networks. Not that Hughes in any way talks of the infrastructure triggering changes in the superstructure; on the contrary, his networks of power are complete hybrids, though hybrids of a peculiar sort—they lend their nonhuman qualities to what were until then weak, local, and scattered corporate bodies. The management of large masses of electrons, clients, power stations, subsidiaries, meters, and dispatching rooms acquires the formal and universal character of scientific laws.

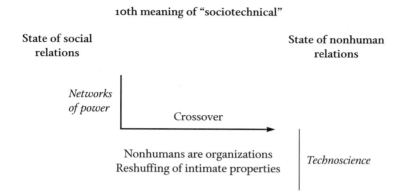

10th meaning of "sociotechnical"

Figure 6.8 Each step in the mythical pragmatogony may be sketched as a crossover through which skills and properties learned in social relations are made relevant for establishing relations within nonhumans. By convention, the next step will be understood as going in the opposite direction.

This ninth layer of meaning resembles the eleventh, since in both cases the crossover goes roughly from nonhumans to corporate bodies. (What can be done with electrons can be done with electors.) But the intimacy of human and nonhuman is less apparent in networks of power than in political ecology. Edison, Bell, and Ford mobilized entities that looked like matter, that seemed nonsocial, whereas political ecology involves the fate of nonhumans already socialized, so closely related to us that they have to be protected by delineation of their legal rights.

Level 8: Industry

Philosophers and sociologists of techniques tend to imagine that there is no difficulty in defining material entities because they are objective, unproblematically composed of forces, elements, atoms. Only the social, the human realm, is difficult to interpret, we often think, because it is complexly historical and, as they say, "symbolic." But whenever we talk of matter we are really considering, as I am trying to show here, a *package* of former crossovers between social and natural elements, so that what we take to be primitive and pure terms are belated and mixed ones. Already we have seen that matter varies greatly from layer to layer—matter in the layer I have called "political ecology" differs from that in the layers called "technology" and "networks of power." Far from being primitive, immutable, and ahistorical, matter too has a complex genealogy and is handed down to us through a long and convoluted pragmatogony.

The extraordinary feat of what I will call *industry* is to extend to matter a further property that we think of as exclusively social, the capacity to relate to others of one's kind, to conspecifics, so to speak. Nonhumans have this capacity when they are made part of the assembly of actants that we call a machine: an automaton endowed with autonomy of some sort and submitted to regular laws that can be measured with instruments and accounting procedures. From tools held in the hands of human workers, the shift historically was to assemblies of machines, where tools relate to one another, creating a massive array of labor and material relations in factories that Marx described as so many circles of hell. The paradox of this stage of relations between humans and nonhumans is that it has been termed "alienation," dehumanization, as if this were the first time that poor and exploited human weakness was confronted by an all-powerful objective force. However, to relate nonhumans together in an assembly of machines, ruled by laws and accounted for by instruments, is to grant them a sort of social life.

Indeed, the modernist project consists in creating this peculiar hybrid: a fabricated nonhuman that has nothing of the character of society and politics yet builds the body politic all the more effectively because it seems completely estranged from humanity. This famous shapeless matter, celebrated so fervently throughout the eighteenth and nineteenth centuries, which is there for Man's—but rarely Woman's—ingenuity to mold and fashion, is only one of many ways to socialize nonhumans. They have been socialized to such an extent that they now have the capacity to create an assembly of their own, an automaton, checking and surveying, pushing and triggering other automata, as if with full autonomy. In effect, however, the properties of the "megamachine" (see Level 7) have been extended to nonhumans.

It is only because we have not undertaken an anthropology of our modern world that we can overlook the strange and hybrid quality of matter as it is seized and implemented by industry. We take matter as mechanistic, forgetting that mechanism is one half of the modern definition of society*. A society of machines? Yes, the eighth meaning of the word sociotechnical, though it seems to designate an unproblematic industry, dominating matter through machinery, is the strangest sociotechnical imbroglio yet. Matter is not a given but a recent historical creation.

Level 7: The Megamachine

But where does industry come from? It is neither a given nor the sudden discovery by capitalism of the objective laws of matter. We have to imagine its genealogy through earlier and more primitive meanings of the term sociotechnical. Lewis Mumford has made the intriguing suggestion that the megamachine—the organization of large numbers of humans via chains of command, deliberate planning, and accounting procedures—represents a change of scale that had to be made before wheels and gears could be developed (Mumford 1966). At some point in history human interactions come to be mediated through a large, stratified, externalized body politic that keeps track, through a range of "intellectual techniques" (writing and counting, basically), of the many nested subprograms for action. When some, though not all, of these subprograms are replaced by nonhumans, machinery and factories are born. The nonhumans, in this view, enter an organization that is already in place and take on a role rehearsed for centuries by obedient human servants enrolled in the imperial megamachine.

In this seventh level, the mass of nonhumans assembled in cities by an internalized ecology (I will define this expression shortly) has been brought to bear on empire building. Mumford's hypothesis is debatable, to say the least, when our context of discussion is the history of technology; but the hypothesis makes excellent sense in the context of my pragmatogony. Before it is possible to delegate action to nonhumans, and possible to relate nonhumans to one another in an automaton, it must first be possible to nest a range of subprograms for action into one another without losing track of them. Management, Mumford would say, precedes the expansion of material techniques. More in keeping with the logic of my story, one might say that *whenever we learn something about the management of humans, we shift that knowledge to nonhumans and endow them with more and more organizational properties*. The even-numbered episodes I have recounted so far follow this pattern: industry shifts to nonhumans the management of people learned in the imperial machine, much as technoscience shifts to nonhumans the large-scale management learned through networks of power. In the odd-numbered levels, the opposite process is at work: *what has been learned from nonhumans is reimported so as to reconfigure people.*

Level 6: Internalized Ecology

In the context of layer seven, the megamachine seems a pure and even final form, composed entirely of social relations; but, as we reach layer six and examine what underlies the megamachine, we find the most extraordinary extension of social relations to nonhumans: agriculture and the domestication of animals. The intense socialization, reeducation, and reconfiguration of plants and animals—so intense that they change shape, function, and often genetic makeup—is what I mean by the term "internalized ecology." As with our other even-numbered levels, domestication cannot be described as a sudden access to an objective material realm that exists *beyond* the narrow limits of the social. In order to enroll animals, plants, proteins in the emerging collective, one must first endow them with the social characteristics necessary for their integration. This shift of characteristics results in a manmade landscape for society (villages and cities) that completely alters what was until then meant by social and material life. In describing the sixth level we may speak of urban life, empires, and organizations, but not of society and techniques—or of symbolic representation and infrastructure. So profound are the changes entailed at this level that we pass beyond the gates of history and enter more profoundly those of prehistory, of mythology.

Level 5: Society

What is a society, the starting point of all social explanations, the *a priori* of all social science? If my pragmatogony is even vaguely suggestive, society cannot be part of our final vocabulary, since the term had itself to be made—"socially constructed" as the misleading expression goes. But according to the Durkheimian interpretation, a society is primitive indeed: it precedes individual action, lasts very much longer than any interaction does, dominates our lives; it is that in which we are born, live, and die. It is externalized, reified, more real than ourselves, and hence the origin of all religion and sacred ritual, which for Durkheim are nothing but the return, through figuration and myth, of the transcendent to individual interactions.

And yet society itself is constructed only through such quotidian interactions. However advanced, differentiated, and disciplined society becomes, we still repair the social fabric out of our own, immanent knowledge and methods. Durkheim may be right, but so is Harold Garfinkel. Perhaps the solution, in keeping with the generative principle of my genealogy, is to look for nonhumans. (This explicit principle is: look for nonhumans when the emergence of a social feature is inexplicable; look to the state of social relations when a new and inexplicable type of object enters the collective.) What Durkheim mistook for the effect of a *sui generis* social order is simply the effect of having brought so many techniques to bear on our social relations. It was from techniques, that is, the ability to nest several subprograms, that we learned what it means to subsist and expand, to accept a role and discharge a function. By reimporting this competence into the definition of society, we taught ourselves to reify it, to make society stand independent of fast-moving interactions. We even learned how to delegate to society the task of relegating us to roles and functions. Society exists, in other words, *but is not socially constructed*. Nonhumans proliferate below the bottom line of social theory.

Level 4: Techniques

By this stage in our speculative genealogy we can no longer speak of humans, of anatomically modern humans, but only of social prehumans. At last we are in a position to define technique, in the sense of a *modus operandi*, with some precision. Techniques, we learn from archaeologists, are articulated subprograms for actions that subsist (in time) and extend (in space). Techniques imply not society (that late-developing hybrid) but a semisocial organization that brings together nonhumans from very different seasons, places, and materials. A bow and arrow, a javelin, a hammer, a net, an article of clothing are composed of parts and pieces that require recombination in sequences of time and space that bear no relation to their original settings. Techniques are what happen to tools and nonhuman actants when they are processed through an organization that extracts, recombines, and socializes them. Even the simplest techniques are sociotechnical; even at this primitive level of meaning, forms of organization are inseparable from technical gestures.

Level 3: Social Complication

But what form of organization can explain these recombinations? Recall that at this stage there is no society, no overarching framework, no dispatcher of roles and functions; there are merely interactions among prehumans. Shirley Strum and I call this third layer of meaning *social complication* (Strum and Latour 1987). Here complex interactions are marked and followed by nonhumans enrolled for a specific purpose. What purpose? Nonhumans stabilize social negotiations. Nonhumans are at once pliable and durable; they can be shaped very quickly but, once shaped, last far longer than the interactions that fabricated them. Social interactions are extremely labile and transitory. More precisely, either they are negotiable but transient or, if they are encoded (for instance) in the genetic makeup, they are extremely durable but difficult to renegotiate. The involvement of nonhumans resolves the contradiction between durability and negotiability. It becomes possible to follow (or "blackbox") interactions, to recombine highly complicated tasks, to nest subprograms into one another. What was impossible for complex* social animals to accomplish becomes possible for prehumans—who use tools not to acquire food but to fix, underline, materialize, and keep track of the social realm. Though composed only of interactions, the social realm becomes visible and attains through the enlistment of nonhumans—tools—some measure of durability.

Level 2: The Basic Tool Kit

The tools themselves, wherever they came from, offer the only testimony on behalf of hundreds of thousands of years. Many archaeologists proceed on the assumption that the basic tool kit (as I call it) and techniques are directly related by an evolution of tools into composite tools. But there is no *direct* route from flints to nuclear power plants. Further, there is no direct route, as many social theorists presume there to be, from social complication to society, megamachines, networks. Finally, there is not a set of parallel histories, the history of infrastructure and the history of superstructure, but only one sociotechnical history (Latour and Lemonnier 1994).

What, then, is a tool? The extension of social skills to nonhumans. Machiavellian monkeys and apes possess little in the way of techniques, but can devise social tools (as Hans Kummer has called them; Kummer 1993) through complex strategies of manipulating and modifying one another. If you grant the prehumans of my own mythology the same kind of social complexity, you grant as well that they may generate tools by *shifting* that competence to nonhumans, by treating a stone, say, as a social partner, modifying it, then using it to act on a second stone. Prehuman tools, in contrast to the ad hoc implements of other primates, also represent the extension of a skill rehearsed in the realm of social interactions.

Level 1: Social Complexity

We have finally reached the level of the Machiavellian primates, the last circumvolution in Daedalus's maze. Here they engage in social interactions to repair a constantly decaying social order. They manipulate one another to survive in groups, with each group of conspecifics in a state of constant mutual interference (Strum 1987). We call this state, this level, social complexity. I will leave it to the ample literature of primatology to show that this stage is no more free of contact with tools and techniques than any of the later stages (McGrew 1992).

An Impossible but Necessary Recapitulation

I know I should not do it. I more than anyone ought to see that it is madness, not only to peel away the different meanings of sociotechnical, but also to recapitulate all of them in a single diagram, as if we could read off the history of the world at a glance. And yet it is always surprising to see how few alternatives we have to the grandiose scenography of progress. We may tell a lugubrious countertale of decay and decadence as if, at each step in the extension of science and technology, we were stepping down, away from our humanity. This is what Heidegger did, and his account has the somber and powerful appeal of all tales of decadence. We may also abstain from telling any master narrative, under the pretext that things are always local, historical, contingent, complex, multiperspectival, and that it is a crime to hold them all in one pathetically poor scheme. But this ban on master narratives is never very effective, because, in the back of our minds, no matter how firmly we are convinced of the radical multiplicity of existence, something surreptitiously gathers everything into one little bundle which may be even cruder than my diagrams—including the postmodern scenography of multiplicity and perspective. This is why, against the ban on master narratives, I cling to the right to tell a "servant" narrative. My aim is not to be reasonable, respectable, or sensible. It is to fight modernism by finding the hideout in which science has been held since being kidnapped for political purposes I do not share.

If we gather in one table the different layers I have briefly outlined— one of my other excuses is how brief the survey, covering so many millions of years, has been!—we may give some sense to a story in which the further we go the more articulated are the collectives we live in (see Figure 6.9). To be sure, we are not ascending toward a future made of more subjectivity and more objectivity. But neither are we descending, chased ever further from the Eden of humanity and *poesis*.

Even if the speculative theory I have outlined is entirely false, it shows, at the very least, the possibility of imagining a genealogical alternative to the dualist paradigm. We are not forever trapped in a boring alternation between objects or matter and subjects or symbols. We are not limited to "not only . . . but also" explanations. My little origin myth makes apparent the impossibility of having an artifact that does not incorporate social relations, as well as the impossibility of defining social structures without accounting for the large role played in them by nonhumans.

Second, and more important, the genealogy demonstrates that it is false to claim, as so many do, that once we abandon the dichotomy between society and techniques we are faced with a seamless web of factors in which all is included in all. The properties of humans and nonhumans cannot be swapped haphazardly. Not only is there an order in the exchange of properties, but in each of the eleven layers the meaning of the word "sociotechnical" is clarified if we consider the exchange: that which has been learned from nonhumans and reimported

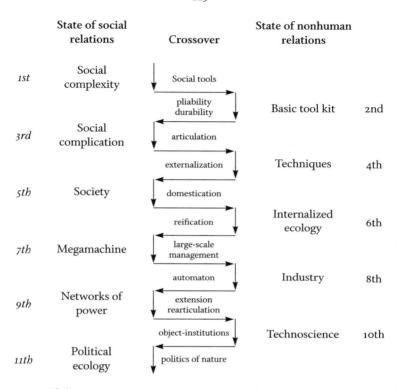

	State of social relations	Crossover	State of nonhuman relations	
1st	Social complexity	Social tools		
		pliability durability	Basic tool kit	2nd
3rd	Social complication	articulation		
		externalization	Techniques	4th
5th	Society	domestication		
		reification	Internalized ecology	6th
7th	Megamachine	large-scale management		
		automaton	Industry	8th
9th	Networks of power	extension rearticulation		
		object-institutions	Technoscience	10th
11th	Political ecology	politics of nature		

Figure 6.9 If the successive crossovers are summed up, a pattern emerges: relations among humans are made out of a previous set of relations that related nonhumans to one another; these new skills and properties are then reused to pattern new types of relations among nonhumans, and so on; at each (mythical) stage the scale and the entanglement increase. The key feature of this myth, is that, at the final stage, the definitions we can make of humans and nonhumans should recapitulate all the earlier layers of history. The further we go, the less pure are the definitions of humans and nonhumans.

into the social realm, that which has been rehearsed in the social realm and exported back to the nonhumans. Nonhumans too have a history. They are not material objects or constraints. Sociotechnical[1] is different from sociotechnical[6] or[7] or[8] or[11]. By adding superscripts we are able to qualify the meanings of a term that until now has been hopelessly confused. In place of the great vertical dichotomy between society and techniques, there is conceivable (in fact, now, available) a range of horizontal distinctions between very different meanings of the sociotechnical hybrids. It is possible to have our cake and eat it too—to be monists and make distinctions.

All this is not to claim that the old dualism, the previous paradigm, had nothing to say for itself. We do indeed alternate between states of social and states of nonhuman relations, but this is not the same as alternating between humanity and objectivity. The mistake of the dualist paradigm was its definition of humanity. Even the shape of humans, our very body, is composed to a great extent of sociotechnical negotiations and artifacts. To conceive of humanity and technology as polar opposites is, in effect, to wish away humanity: we are sociotechnical animals, and each human interaction is sociotechnical. We are never limited to social ties. We are never faced only with objects. This final diagram relocates humanity right where we belong—in the crossover, the central column, the articulation, the possibility of mediating between mediators.

But my main point is that, in each of the eleven episodes I have traced, an increasingly large number of humans are mixed with an increasingly large number of nonhumans, to the point that, today, the whole planet is engaged in the making of politics, law, and soon, I suspect, morality. The illusion of modernity was to believe that the more we grew, the more separate objectivity and subjectivity would become, thus creating a future radically different from our past. After the paradigm shift in our conception of science and technology, we now know that this will never be the case, indeed that this has never *been* the case. Objectivity and subjectivity are not opposed, they grow together, and they do so irreversibly. At the very least, I hope I have convinced the reader that, if we are to meet our challenge, we will not meet it by considering artifacts as things. They deserve better. They deserve to be housed in our intellectual culture as full-fledged social actors. Do they mediate our actions? No, they are us. The goal of our philosophy, social theory, and morality is to invent political institutions that can absorb this much history, this vast spiraling movement, this labyrinth, this fate.

The nasty problem we now have to deal with is that, unfortunately, we do *not* have a definition of politics that can answer the specifications of this nonmodern history. On the contrary, every single definition we have of politics comes from the modernist settlement and from the polemical definition of science that we have found so wanting. Every one of the weapons used in the science wars, including *the very distinction* between science and politics, has been handed down to the combatants by the side we want to oppose. No wonder we always lose and are accused of politicizing science! It is not only the practice of science and technology that epistemology has rendered opaque, but also that of politics. As we shall soon see, the fear of mob rule, the proverbial scenography of might versus right, is what holds the old settlement together, is what has rendered us modern, is what has kidnapped the practice of science, all for the most implausible political project: that of doing away with politics.

The Invention of the Science Wars

The Settlement of Socrates and Callicles

"If Right cannot prevail, then Might will take over!" How often have we heard this cry of despair? How sensible it is to cry for Reason in this way when faced with the horrors we witness every day. And yet this cry too has a history, a history that I want to probe because doing so may allow us to distinguish science from politics once again and maybe to explain why the Body Politic has been invented in such a way as to be rendered impossible, impotent, illegitimate, a born bastard.

When I say that this rallying cry has a history, I do not mean that it moves at a fast pace. On the contrary, centuries may pass without affecting it a bit. Its tempo resembles that of Fermat's theorem, or plate tectonics, or glaciations. Witness for instance the similarity between Socrates' vehement address to the Sophist Callicles, in the famous dialogue of the *Gorgias,* and this recent instance by Steven Weinberg in the *New York Review of Books:*

> Our civilization has been powerfully affected by the discovery that nature is strictly governed by impersonal laws . . . We will need to confirm and strengthen the vision of a rationally understandable world if we are to protect ourselves from the irrational tendencies that still beset humanity. (August 8, 1996, 15)

And here is Socrates' famed admonition: *geômetrias gar ameleis!*

In fact, Callicles, the expert's opinion is that co-operation, love, order, discipline, and justice *bind heaven and earth, gods and men.* That's why they call the universe an ordered whole, my friend, rather than a disorderly mess or an unruly shambles. It seems to me that, for all your expertise in the field, you're overlooking this point. You've failed to notice how much power *geometrical equality* has among gods and men, and this *neglect of geometry* has led you to believe that one should try to gain a disproportionate share of things. (507e–508a)

What these two quotations have in common, across the huge gap of centuries, is the strong link they establish between the respect for impersonal natural laws, on the one hand, and the fight against irrationality, immorality, and political disorder, on the other. In both quotations the fate of Reason and the fate of Politics are associated in a single destiny. To attack Reason is to render morality and social peace impossible. Right is what protects us against Might; Reason against civil warfare. The common tenet is that we need something "inhuman"—for Weinberg, the natural laws no human has constructed; for Socrates, geometry whose demonstrations escape human whim—if we want to be able to fight against "inhumanity." To sum up: only inhumanity will quash inhumanity. Only a Science that is not made by man will protect a Body Politic that is in constant risk of being made by the mob. Yes, Reason is our rampart, our Great Wall of China, our Maginot Line against the dangerous unruly mob.

This line of reasoning, which I will call "inhumanity against inhumanity," has been attacked ever since it began, from the Sophists, against whom Plato launches his all-out assault, all the way to the motley gang of people accused of "postmodernism" (an accusation, by the way, as vague as the curse of being a "sophist"). Postmoderns of the past and of the present have tried to break the connection between the discovery of natural laws of the cosmos and the problems of making the Body Politic safe for its citizens. Some have claimed that adding inhumanity to inhumanity has simply increased the misery and the civil strife and that a staunch fight against Science and Reason should be started to protect politics against the intrusion of science and technology. Still others, who are targeted publicly today and with whom, I am sorry to say, I am often lumped by mistake, have tried to show that mob rule, the violence of the Body Politic, is everywhere polluting the purity of Science, which becomes every day more human, all too human, and every day more adulterated by the civil strife it was supposed to assuage. Others, like Nietzsche, have shamelessly accepted Callicles' position and claimed, against the degenerate and moralistic

Socrates, that only violence could bend both the mob and its retinue of priests and other men of *ressentiment,* among whom, I am sorry to say, he included scientists and cosmologists like Weinberg.

None of these critiques, however, has disputed *simultaneously* the definition of Science *and* the definition of the Body Politic that it implies. Inhumanity is accepted in both or in at least one of them. Only the connection between the two, or its expediency, has been disputed. In this chapter and the next I want to go back to the source of what I call the scenography of the fight of Right against Might, to see how it was staged in the first place. I want, in other words, to attempt the archaeology of the Pavlovian reflex that makes any lecture in science studies trigger these questions from the audience: "Then you want force alone to decide in matters of proof? Then you are for mob rule against that of rational understanding?" Is there really no other way? Is it really impossible to build up other reflexes, other intellectual resources?

To go some way toward this genealogy, no text is more appropriate than the *Gorgias,* especially in the lively translation by Robin Waterfield (Oxford University Press, 1994), since never was the genealogy more beautifully set up than in the acrimonious debate between Socrates and Callicles, which has been commented on by all the later Sophists from Greece and then from Rome, as well as, in our time, by thinkers as different as Charles Perelman and Hannah Arendt. I am not reading the *Gorgias* as if I were a Greek scholar (I am not, as will become painfully clear) but as if it had been published a few months ago in the *New York Review of Books* as a contribution to the raging Science Wars. Fresh as in 385 B.C., it deals with the same puzzle as the one besetting the academy and our contemporary societies today.

This puzzle can be stated very simply: the Greeks made one invention too many! They invented both democracy and mathematical demonstration, or to use the terms Barbara Cassin comments on so beautifully, *epideixis** and *apodeixis** (Cassin 1995). We are still struggling, in our "mad cow times," with this same quandary, how to have a science *and* a democracy together. What I call the settlement between Socrates and Callicles has made the Body Politic incapable of swallowing the two inventions at once. More fortunate than the Greeks, we may be able, if we rewrite this settlement, to profit at last from both.

To revisit this "primal scene" of Might and Right, I am afraid we have to follow the dialogue in some detail. The structure of the story is clear. Three Sophists in turn oppose Socrates and are defeated one after the other: Gorgias, a bit tired from a lecture he just gave; Polus, a bit slow; and finally the harshest of the three, the famous and infamous Callicles. At the end, Socrates, having discouraged discussion, speaks to himself and makes a final appeal to the shadows of the afterworld, the only ones able to understand his position and to judge it—with good reason, as we shall see.

In my commentary I will not always follow the chronological order of the dialogue and will focus mainly on Callicles. I want to point out two features of the discussion that, in my view, have often been overlooked. One is that Socrates and his third opponent, Callicles, agree on everything. Socrates' invocation of reason against the unreasonable people is actually patterned on Callicles' request for "an unequal share of power." The second feature is that it is still possible to recognize in the four protagonists' speeches the dim trace of the *conditions of felicity** that are proper to politics and that both Callicles and Socrates (as characters in Plato's puppet show at least) have tried their best to erase. This will be the focus of Chapter 8, in which I will try to show that the Body Politic could behave very differently if another definition of science and of democracy were provided. A science freed at last from its kidnapping by politics? Even better, a polity freed at last from its delegitimation by science? It is certainly, everyone would admit, worth a try.

Socrates and Callicles versus the People of Athens

The Demotic Hatred

We are so used to opposing Might and Right and to looking in the *Gorgias* for their best instantiations that we forget to notice that Socrates and Callicles have a common enemy: the people of Athens, the crowd assembled in the agora, talking endlessly, making the laws at their whim, behaving like children, like sick people, like animals, shifting opinions whenever the wind changes direction. Socrates accuses Gorgias and then Polus of being the slaves of the people, or of being like Callicles, unable to utter other words than those the raging crowd puts in his mouth. But Callicles too, when it is his turn to talk, accuses Socrates of being enslaved by the people of Athens and of for-getting what makes noble masters superior to the *hoi polloi:* "You pre-tend that truth is your goal, Socrates, but in actual fact you steer discussions towards this kind of ethical idea—ideas which are *unso-phisticated* enough to have *popular appeal,* and which depend entirely on convention, not on nature" (482e).

The two protagonists compete in trying to avoid being branded with that fatal accusation: *resembling* the people, the common people, the menial manual people of Athens. As we will see, they soon disagree on how best to break the majority rule, but the goal of breaking the rule of the crowd remains beyond question. Witness this exchange, in which a condescending and tired Callicles seems to lose the contest over how much *distance* one should keep from the *demos:*

> CALLICLES: I can't explain it, Socrates, but I do think you're making your points well. All the same, I'm feeling what people invariably feel with you: I'm not *entirely* convinced.
> SOCRATES: It's the *demotic love* residing in your heart which is resisting me, Callicles.(513c)

Obviously the love of the people is not stifling Socrates' breath! He has a way to break the rule of majority that no obstacle can restrain. What should we call what resists in *his* heart if not "demotic hatred"? If you make a list of all the derogatory terms with which the common crowd is branded by Callicles and Socrates, it is hard to see which of them despises it most. Is it because assemblies are polluted by women, children, and slaves that they deserve this scorn? Is it because they are made up of people who work with their hands? Or is it because they switch opinions like babies and want to be spoiled and overfed like ir-responsible children? All of that, to be sure, but their worst quality, for our two protagonists, is even more elementary: the great constitutive defect of the people is that there are simply *too many* of them. "A rhetorician, then," says Socrates with his tranquil arrogance, "isn't concerned to educate the people *assembled* in lawcourts and so on about right and wrong; all he wants to do is *persuade* them. I mean, I shouldn't think it's possible for him to get *so many people* to under-stand such *important matters in such a short time*" (455a).

Yes, there are too many of them, the questions are too important [*megala pragmata*], there is too little time [*oligô chronô*]. Are these not, however, the normal conditions of the Body Politic? Is it not to deal with these peculiar situations of number, urgency, and priority that the subtle skills of politics were invented? Yes, as we shall see in Chapter 8, but this is *not* the tack that Socrates and Callicles take. Horror-struck by the numbers, the urgency, and the importance, they agree on another radical solution: break the majority rule and escape from it. It is at this juncture that the fight between Might and Right is being invented, the *commedia dell'arte* scenography that is going to entertain so many people for so long.

Because of the clever staging by Plato (so clever that it continues even today in the campus amphitheatres) we have to distinguish between two roles played by Callicles, so that we don't attribute to the Sophists the position in which Socrates is trying to corner them—a position they kindly accept because Plato is holding all the puppet strings of the dialogue at once. Believing what Plato says of the Sophists would be like reconstituting science studies from the science warriors' pamphlets! I will thus call the Callicles playing the role of a foil for Socrates the *straw Callicles*. The Callicles that retains features of the precise conditions of felicity invented by the Sophists and still visible in the dialogue, I will call the *positive*, or the *historical*, or the *anthropological* Callicles. While the straw Callicles is a strong enemy of the demos and the perfect counterpart for Socrates, the anthropological Callicles will allow us to retrieve some of the specificities of political truth-saying.

How Best to Break the Majority Rule

Callicles' solution is well known. It is the age-old aristocratic solution, presented in a crisp and naive light by the Nietzschean blond brute descended from a race of masters. But we should not be taken in by what happens on the stage. Callicles is not for Might understood as "mere force" but for something, on the contrary, that will make might weak. He is looking for a might mightier than might. We should follow with some precision the tricks that Callicles employs, because, in spite of his sneering remarks, it is on the bad guy that the good guy, Socrates, is going to pattern *his* copycat solution to the *same* problem: for both, *beyond* the conventional laws made for and by the mob, there is another natural law reserved for the elite, which makes the noble souls unaccountable to the demos.

In a visionary anticipation of certain aspects of sociobiology, Callicles appeals to nature above manmade history:

> But I think we only have to look at *nature* to find evidence that it is *right* for *better* to have a *greater* share than worse, more capable than less capable. The evidence for this is widespread. Other *creatures* show, as do human communities and nations, that *right* has been determined as follows: the *superior* person shall *dominate* the inferior person and have more than him . . . These people act, surely, in conformity with the natural essence *[kata phusin]* of right and, yes, I'd even go so far as to say that they act in conformity with *natural law [kata nomon ge tès phuseôs]*, even though they presumably *contravene* our *man-made* laws. (483c–e)

As Socrates and Callicles immediately see, however, this is not a sufficient definition of Might, for a simple and paradoxical reason: Callicles who appeals to the superior natural law is nonetheless physically *weaker* than the crowd. "Presumably you don't think that two people are better than one, or that your slaves are *better* than you, just because they're *stronger* than you" (489d), says Socrates ironically. "Of course," says Callicles, "I mean that *superior* people are *better*. Haven't I been telling you all along that 'better' and 'superior' are the same, in my opinion? What else do you think I've been saying? That law consists of the statements made by an *assembly of slaves and assorted other forms of human debris* who could be completely *discounted* if it weren't for the fact that they do have *physical strength* at their disposal" (489c).

We should be careful here not to introduce the moral argument that will come later, and we should focus only on Callicles' way of escaping the rule of the majority. His appeal to irrepressible natural law exactly resembles the "inhumanity to quash inhumanity" with which I started this chapter. Stripped of its moral dimension, which will be added later in the dialogue in the interests of staging, not of logic, Callicles' plea becomes a moving appeal to a force stronger than the democratic force of the assembled people, a force beautifully defined by Socrates when he summarizes Callicles' position:

> SOCRATES: Here's your position, then: a *single* clever person is almost *bound to be superior to ten thousand fools;* political power should be his

and they should be his subjects; and it is appropriate for someone with political power to have *more* than his subjects. Now, I'm not picking on the form of words you used, but that, I take it, is the implication of what you're saying—of a single individual being *superior* to *ten thousand others.*

CALLICLES: Yes, that's what I mean. In my opinion, that's what natural right is—for an individual who is *better* (that is, more clever) to *rule over* second-rate people and to have *more* than them. (490a)

Thus when Might enters the scene in the person of the Nietzschean Callicles, it is not as the Brownshirts smashing their way through the laboratories—as in the nightmares epistemologists have when they think of science studies—it is as an elitist and specialized expertise breaking the neck of mob rule and imposing a Right superior to all the conventional property rights. When Might is invoked on the stage it is not as a crowd against Reason, it is as *one* man against the crowd, against myriad fools. Nietzsche has deftly drawn the moral of this paradox in his famous advice: "One should always defend the strong against the weak." Nothing is more elitist than the nightmarish Might.

The model employed by Callicles is of course nobility, the aristocratic upbringing to which Plato himself, as has been so often noticed, owes his virtue. Nobility gives an ingrained quality and a native status that makes masters different from the hoi polloi. But Callicles shifts the classic pattern considerably by complementing upbringing with an appeal to a law that is superior to the law. Elites are defined not only by their past and their ancestors but also by their connection to this natural law that does not depend on the "social construction" made by slaves. We are so used to laughing when Callicles falls into all the traps set by Socrates that we fail to see how similar are the roles both offer to an irrepressible natural law that is not manmade: "What do we do with the best and strongest among us?" asks Callicles.

We capture them young, like lions, mould them, and *turn them into slaves* by chanting spells and incantations over them which insist that they have to be equal to others and that equality is admirable and right. But I'm sure that if a man is born *in whom nature* is *strong* enough, he'll *shake off all these limitations,* shatter them to pieces, and win his freedom; he'll *trample* all our regulations, charms, spells, and *unnatural* laws into the dust; this slave will rise up and reveal himself as our master; and then *natural right [to tès phuseôs dikaion]* will blaze forth. (483e–484b)

This sort of sentence has done a lot for Callicles' bad reputation, and yet it is the same irrepressible urge that even bad education cannot spoil will "shake off" irrationality and "blaze forth" when Socrates defeats *his* ten thousand fools. If you remove from Callicles the cloak of immorality, if you make him swap offstage his brutish and hairy wig for the virginal white cloth of Antigone, we will be forced to notice that his plea possesses the same beauty as hers against Creon, over which so many moral philosophers have shed so many tears. Both say that deformation by the "social construction" cannot stop the natural law from "blazing forth" in the hearts of naturally good people. In the long run, the noble hearts will triumph over manmade conventions. We despise the Callicleses and we praise the Socrateses and the Antigones, but this is to hide the simple fact that they all wish to stand alone against the people. We complain that without Right the war of all against all will take over, but we fail to notice *this* war of two, Socrates and Callicles, against all the others.

With this little warning in mind, we can now listen to Socrates' solution with a different ear. On the stage, to be sure, he has a field day when ridiculing Callicles' appeal to an unlimited Might: "Would you go back to the beginning, though, and tell me again what you and Pindar mean by natural right? Am I right in remembering that according to you it's the *forcible seizure of property* belonging to inferior people by anyone who is superior, it's the *dominance* of the worse by the better, and it's the *unequal* distribution of goods, so that the élite have more than second-rate people?" (488b).

The entire audience screams in horror when confronted with this threat of Might swallowing the rights of ordinary citizens. But how is Socrates' own solution *technically* different? Again, let the partners stay on the stage for a moment in plain clothes, without the impressive garments of morality, and listen carefully to Socrates' definition of how to resist the *same* assembled crowd. This time it is the poor Polus who suffers the sting of the numbfish:

> The trouble, Polus, is that you're trying to use on me the kind of rhetorical refutation which *people in lawcourts* think is successful. There too, you see, people think they're proving the other side wrong if they produce *a large number of eminent witnesses* in support of the points they're making, but their opponent comes up with only *a sin-*

gle witness or none at all. This kind of refutation, however, is completely *worthless* in the *context of the truth [outos de o elegchos oudenos axios estin pros tèn alètheian]*, since it's perfectly possible for someone to be defeated in court by a *horde of witnesses* with no more than apparent respectability who all testify falsely against him. (471e–472a)

How often his position has been admired! How many voices have quivered in commenting on the courage of one man against the hordes, like Saint Genevieve stopping Attila's throngs with the sheer light of her virtue! Yes, it is admirable, but no more than Callicles' appeal to a natural law. The goal is the same, and even Callicles, in his wildest definition of forceful domination, never dreams of a position of power as dominant, as exclusive, as undisputed as the one Socrates requests for his knowledge. It is a great power to which Socrates appeals, comparing it to the physician's knowledge of the human body since it can enslave all the other forms of expertise and know-how: "They don't realize that this kind of expertise should properly be the *dominant* kind, and should be allowed *a free hand* with the products of all those other techniques because it knows—and none of the others does—which food and drink promotes a good physical state and which doesn't. That's why the *rest of them* are suited only for *slavish, ancillary, and degrading work,* and should *by rights* be subordinate to training and medicine" (517e–518a).

Truth enters and the agora is emptied. One man can triumph over everyone else. In the "context of truth," as in the "context of aristocracy," the hordes are defeated by a force—yes a force—superior to the reputation and physical force of the demos and to its endless and useless practical knowledge. When Might comes onstage, as I said above, it is not as a crowd but as one man *against* the crowd. When Truth enters the scene, it is not as one man against everyone else, it is as an impersonal, transcendent natural law, a Might mightier than Might. Arguments prevail against everything else because they are rationally made. This is what Callicles has missed: the power of geometrical equality: "You neglected geometry, Callicles!" The young man will never recover from the blow.

That Callicles and Socrates are acting like Siamese twins in this dialogue is made explicit by Plato's many parallels between his heroes' two solutions. Socrates compares Callicles' slavish attachment to the demos with his own slavish attachment to philosophy: "I love Alcibiades the son of Cleinias, and philosophy, and your two loves are the Athenian *populace* and Demus the son of Pyrilampes . . . So rather than expressing surprise at the things I've been saying, you should stop my darling philosophy voicing these opinions. You see, my friend, she is *constantly repeating* the views you've just heard from me, and she's far *less fickle* than my other love. I mean, Alcibiades says different things at different times, but *philosophy's views never change*" (481d–482a).

Against the capricious people of Athens, against the even more whimsical Alcibiades, Socrates has found an anchor that allows him to be right against everyone else's vagaries. But this is also, in spite of Socrates' sneering remark, what Callicles thinks of natural laws: they protect him against the vagaries of the assembled people. There is, to be sure, a big difference between the two anchors, but this should count in favor of the real anthropological Callicles, not Socrates: the good guy's anchor is fastened in the ethereal afterworld of shadows and phantoms, whereas Callicles' anchor is at least gripping the solid and resisting matter of the Body Politic. Which one of the two anchors is better secured? Incredible as it seems, Plato manages to make us believe that it is Socrates'!

The beauty of the dialogue, as has often been noticed, lies mainly in the opposition between two parallel scenes, one in which Callicles mocks Socrates for being unable to defend himself in the tribunal of *this* world, and the other at the end, when Socrates mocks Callicles for being unable to defend himself in the afterworld tribunal of Hades. Round one:

> Socrates, you're neglecting matters you shouldn't neglect. Look at the noble temperament with which nature has endowed you! Yet what you're famous for is behaving like a teenager. You couldn't deliver a *proper* speech *to the councils* which administer justice, or make a *plausible* and *persuasive* appeal . . . The point is that if you or any of your sort were seized and taken away to prison, unjustly accused of some crime, you'd be *incapable*—as I'm sure you're well aware—of doing anything for your self. *With your head spinning and mouth gaping open,* you wouldn't know what to say. (485e–486b)

A terrible situation indeed for a Greek to be left speechless by an unfair accusation in the midst of the crowd. Notice that Callicles does not admonish Socrates for being too lofty, but for being an impotent, lowly, and idiotic teenager. Callicles has a resource of his own that comes from an ancient aristocratic tradition: an inherited talent for speech which allows him to find just the right thing to say against the conventions created by "second-rate citizens."

To find a retort to that one, Socrates has to wait until the end of the dialogue, and must abandon his dialectic of questions and answers to tell a crepuscular tale. The final round:

> I think *it's a flaw in you* that you won't be able to defend yourself when the time comes for you to undergo the trial and the assessment which I've just been talking about. Instead, when you come to be judged by that son of Aegina [Rhadamanthys] and he seizes you and takes you away, *your head will spin and your mouth will gape* there in *that* world just as much as mine would *here*, and the chances are that someone will smash you in the face and generally abuse you as if you were a *nobody* without any *status* at all. (526e–527a)

A beautiful effect on the stage, to be sure, with naked shadows pacing a papier-mâché inferno and artificial fumes and fog lingering in the air. "But a bit late, Socrates," the historical and anthropological Callicles could have retorted, "because politics is not about the naked dead living in a world of phantoms and judged by half-existing sons of Zeus, but about clothed and living bodies assembled in the agora with their status and their friends, in the bright sun of Attica, and trying to decide, on the spot, in real time, what to do next." But the straw Callicles, by now, through a happy coincidence, has been shut down by Plato. So much for the dialectical method and the appeal to "the community of free speech." When the time of retribution has come, Socrates speaks alone in the much despised epideictic way (465e).

It is a pity that the dialogue ends with such an admirable but empty appeal to the shadows of politics, because Callicles could have shown that even his selfish and extravagant claim to hedonism, which made him so contemptible to the theater crowd, is also used by Socrates to define *his* way of dealing with the people: "And yet, my friend, in my opinion, it's preferable for me to be a musician with an out-of-tune lyre or a choirleader with a cacophonous choir, and it's preferable for *almost everyone in the world* to find my beliefs misguided and wrong, rather than for *just one person—me—*to contradict and clash *with myself*" (482b–c).

"Perish the people of Athens," claimed the straw Callicles, "provided I have a good time, and forcibly seize as much as I can from the hands of the second-rate human debris!" In what sense is Socrates' appeal less selfish? "Perish the whole world, provided I am in agreement not only with one other person"—as, we shall see, he has earlier said to Polus—"but with myself!" Knowing that Plato willfully misrepresents Callicles' and Gorgias's position, whereas he presents Socrates as having the last word and responding seriously, who is the more dangerous—the agoraphobic mad scientist, or the "blonde brute of prey"? Who is the more deleterious for democracy, Right or Might? All through the dialogue the parallelism between the solutions of the two sparring partners is inescapable.

And yet it is also completely invisible, as long as we keep our eyes on the stage. Why? Because of the definition of knowledge that Socrates forcibly imposes over Callicles' definition. This is where the symmetry is broken; this is what makes Callicles exit to the sound of boos, no matter how many Nietzscheans will later try to push him back onto the boards. QED; TKO.

The Triangular Contest of Socrates, the Sophists, and the Demos

In the three dialogues of the *Gorgias*, Might and Right never appear as comparable; later we will see why. What remain commensurable enough to be disputed are the relative qualities of two types of expert knowledge: one in the hands of Socrates, the other in the hands of the rhetoricians (a word invented, it seems, in the *Gorgias*). What is beyond question for both Socrates and the straw Sophists is that some expert knowledge is necessary, either to make the people of Athens behave in the right way or to keep them at bay and shut their mouths. They no longer consider the obvious solution to the problem besetting the agora, the one we will explore in Chapter 8, although it is still present in the dialogue, at least as a negative template: the assembled Body Politic, in order to make decisions, *cannot* rely on expert knowledge alone, given the constraints of number, totality, urgency, and priority that politics imposes. Reaching a decision *without* appealing to a natural impersonal law in the hands of experts requires a disseminated knowledge as multifarious as the multitude itself. *The knowledge of the*

whole needs the whole, not the few. But that would be a scandal for Callicles and for Socrates, a scandal whose name has been the same at all periods: democracy.

So here again the disagreement of the partners is secondary to their complete agreement: the contest is about how to shut the mouths of the people faster and tighter. On this ground, Callicles is going to lose fast. After agreeing, with a common paternalism, that experts are needed to "look after a community and its citizens" (513e), the two argue over what sort of knowledge will be best. Rhetoricians have one type of expertise and Socrates has another. One is epideictic, the other apodeictic. One is employed in the dangerous conditions of the agora, the other in the quiet and remote one-to-one conversation Socrates pursues with his disciples. At first glance it looks as if Socrates should lose at this game, since it is of no use at all to have a method for besting the citizens of the agora that is itself agoraphobic and operates only on a one-to-one basis. "I'm content," Socrates confesses naively to Polus, "if *you* testify to the validity of *my* argument, and I canvass *only* for your vote, *without caring about what everyone else thinks*" (476a). But politics is precisely about "caring for what everyone thinks." Canvassing for only one vote is worse than a crime, it is a political mistake. So when Callicles admonishes Socrates for this infantile behavior, he should win the day: "Even a naturally gifted person isn't going to develop into a *real man,* because he's avoiding the *heart of his community* and the *thick of the agora,* which are the places where, as Homer tells us, a man *'earns distinction.'* Instead he spends the rest of his life *sunk out of sight, whispering in a corner* with three or four young men, rather than giving open expression to *important* and *significant* ideas" (485d–e).

Thus the dialogue, logically, should end up with only one scene, in which Socrates is sent back to his campus corner, philosophy being limited to a useless specialized obsession, with no relation to what "real men" do to "earn distinction" with "important and significant ideas." This is what rhetoric will do. But this is not what we did when we reinvented the power of Science, with a capital S, over and over again. With the "context of truth" that Socrates is bringing to the fore, Callicles' triumph becomes impossible. It is a very subtle trick, but it is enough to reverse the logical course of the dialogue and to make Socrates win where he should have lost.

What is the *supplement* provided by apodeictic reasoning that makes it so much better than the natural laws invoked by the Sophists against the conventions of "slaves and assorted human debris"? This kind of reasoning is *beyond dispute:*

> SOCRATES: But can knowledge be either true or false?
> GORGIAS: Certainly not.
> SOCRATES: Obviously, then, *conviction [pistis]* and *knowledge [epistèmè]* are not the same. (454d)

The Sophists' transcendence is beyond convention, but not beyond dispute, since the questions of being superior, more natural, better born, better bred open another swarm of discussions, as can be witnessed even today—no matter how many Bell Curves one throws into the pot. Callicles has invented a way to discount the crowd's physical weight and number, but not to escape altogether from the *site* of the chock-full agora. Socrates' solution is much stronger. The fabulous secret of mathematical demonstration that he has in his hands is that it is a step-by-step persuasion that forces one to assent no matter what. Nothing, though, makes this way of reasoning able to adjust to the extremely harsh conditions of the agora, where it should be as useful, to borrow an old feminist slogan, as a bicycle to a fish. So a bit more work is needed for Socrates to be able to make use of this weapon. He first has to disarm everyone else, or at least make them believe they are thoroughly disarmed: "So we'd better think in terms of *two kinds* of persuasion, one of which confers conviction *without understanding [to men pistin parchomenon aneu tou eidenai],* while the other confers knowledge *[epistèmè]*" (454e).

Epistèmè, how many crimes have been committed in your name! On this the whole history hinges. So venerable is this opposition that, contrary to the obviously rigged fight of Might and Right, we might lose our nerve at this point and fail to see how bizarre and illogical the argument is. The whole difference between the two kinds of persuasion relies on two innocuous little words: "without understanding." But understanding *of what?* If we mean the understanding of the very specific conditions of felicity for political discussion—that is, number, urgency, and priority—then Socrates is certainly wrong. If anything, it is the apodeictic reasoning of causes and consequences, the *epistèmè,* that is "without understanding," meaning that it fails to take into account the pragmatic conditions of deciding what to do next in the thick of the agora with ten thousand people talking all at once. On his own, Socrates cannot replace this pragmatic knowledge *in situ*, with his unsituated knowledge of demonstration. His weapon is mind-boggling, mouth-shutting, but a useless deterrent in the context of the agora. He needs help. Who is going to give him a hand? The foils invented by Plato, who, as usual, conveniently fall into the trap like ideal straw men.

The dialogue could not work and make Socrates triumph against all odds if the puppet Sophists did not share Socrates' loathing for the skills and gimmicks with which common people go about their daily business. So when Socrates makes a distinction between real knowledge and know-how the (straw) Sophists don't protest, since they have the same aristocratic contempt of practice: "There's absolutely no expertise involved in the way it [cookery] pursues pleasure; it hasn't considered either the nature of pleasure or the reason why it occurs . . . All it [the technical cook] can do is remember a *routine* which has become *ingrained* by *habituation* and *past experience*, and that's also what it relies on to provide us with pleasant experiences" (501a–b).

Amusingly enough, this definition of merely practical know-how, although uttered with scorn, would today fit what psychologists, pragmatists, and cognitive anthropologists would call "knowledge." But the key point is that this distinction itself has *no other content* than Socrates' disdain for the common people. Socrates is here on very thin ice. The distinction between knowledge and practical know-how is both what allows him to appeal to a mouth-shutting superior natural law and also what is enforced by the very action of shutting the mouths of the ten thousand people who go about their business every day "without knowing what they do." If they knew what they were doing, the distinction would be lost. So if this absolute demarcation is not imposed by sheer force—the true task of epistemology over the ages—the "context of truth" cannot be brought to bear on the impossibly deleterious atmosphere of public debate. This is one of the rare cases in history in which "sheer force" has been applied. To enforce this divide what do we have? *Only* Socrates' word for it—and the meek retreat of Gorgias, Polus, and Callicles into acceptance of Socrates' definition, carefully staged in Plato's theatrical machinery. That's quite a few conditions for an unconditional appeal to an unconstructed "impersonal law."

As Lyotard showed some time ago, and as Barbara Cassin (Cassin 1995) has recently demonstrated so forcefully, distinguishing the two forms of knowledge and setting up the absolute difference between force and reason requires a *coup de force*—the one that expels the Sophists from philosophy and the common people from rigorous knowledge. Without this coup, the expert knowledge of demonstration could not take over the precise, subtle, necessary, distributed, indispensable knowledge of the members of the Body Politic who take it upon themselves to decide what to do next in the agora. *Epistème* will not replace *pistis*. Apodeictic reasoning will remain important, of course, even indispensable, but *in no way bound to the question of how best to discipline the multitude*. As in the birth of all political regimes, undisputed legitimacy resides in an original bloody coup. In this case, and this is the beauty of the play, the blood that is shed *is Socrates' own*. That sacrifice makes the move even more irresistible and the legitimacy even more indisputable. By the end, there won't be a dry eye left in the theater . . .

The Sophists are no match for this dramatic move, and after accepting, first, that expert knowledge is necessary to replace that of the poor ignorant multitude, and second, that the knowledge of demonstration is absolutely, not relatively, different from the skills and gimmicks of the common people, they have to confess that their form of expertise is empty. How silly Gorgias's boasting now sounds: "Doesn't that simplify things, Socrates? Rhetoric is the only area of expertise you need to learn. You *can ignore all the rest* and still get the *better* of the professionals" (459c).

We will see in the next chapter that this apparently cynical answer is in fact a very precise definition of the *non*professional nature of political action. However, if we agree to overlook this point and if we start to accept the contest and pit the specialized knowledge of scientists against the specialized knowledge of rhetoricians, then sophistry is immediately turned into an empty manipulation. It is like introducing a race car into a marathon; the new machine renders the slower runners ridiculous.

SOCRATES: Faced with phenomena like the one you've mentioned, it comes across as *something supernatural*, with *enormous power*.
GORGIAS: You don't know the half of it, Socrates! Almost every accom-

plishment falls within the scope of rhetoric . . . Often in the past, when I've gone with my brother or some other doctor to one of their patients who was refusing to take his medicine or to let the doctor operate on him or cauterize him, the doctor proved *incapable* of persuading the patient to accept his treatment, but I succeeded, *even though I didn't have any other expertise to draw on except rhetoric.* (456a–b)

Even for sentences like that we need centuries of Pavlovian training to read them as cynical, because what the real Gorgias alludes to here is the impotence of specialists to make the people as a whole make tough decisions. The real Gorgias points out an extraordinarily subtle skill, one that Socrates does not want to understand (although he practices it so cleverly); the puppet Gorgias is made to say that no knowledge at all is necessary. After their staged defeat, the rhetoricians are putting their own heads on the chopping block. Having accepted that rhetoric is an expertise, then having found it empty, they are now expelled from knowledge altogether, and their skills branded as mere "flattery" (502d), one of the many obscure types of popular know-how from which rhetoric cannot be distinguished. "Well, in my opinion, Gorgias, it *doesn't involve* expertise; all you need is a mind which is *good at guessing*, some *courage*, and a natural talent *for interacting with people.* The general term I use to refer to it is 'flattery,' and this strikes me as a multifaceted activity, one of whose branches is *cookery*. And what I'm saying about cookery is that it does seem to be a branch of expertise, but in fact isn't: it is a *knack, acquired by habituation [ouk estin technè, all' empeiria kai tribè]*" (463a–b).

The most moving feature, which will deserve all our attention later, is that even in this famous *coup de grâce* Socrates is still complimenting rhetoric. How can we not consider as positive qualities being "good at guessing," having "courage," knowing "how to interact with people"— certainly not skills that Socrates lacks in spite of his claims to the contrary? For that matter, what is so bad about being as talented as a cook? I myself prefer a good chef to many bad leaders! And yet Socrates has won. The weakest has turned the tables on the strongest. The least logical—that is, the "happy few,"—have won over the "universal" logic, that is, everyone minding the whole Body Politic at once. Socrates, who by his own confession is the least adapted to rule over the people, rules over them—at least from the conveniently far-away place of the Isles of the Blessed: "I think," he says, wrapping his words in three degrees of irony, "I'm the only genuine practitioner of politics in Athens today, the *only example of a true statesman*" (521d).

And it is true: no tyranny has been longer lasting than that by this sacrificed, dead man over the living, no power more absolute, no reign more undisputed.

The defeat of the (straw) Sophists is nothing compared with that of the common people of Athens, as can be seen by a summary of the argument so far. The "human debris and assorted slaves" are the great absent ones, without even a chorus to defend their common sense as in classic tragedies. When we start reading this most famous dialogue carefully, we discover not only a fight between Callicles (that is, Might) and Socrates (Right), but *two* overlapping disputes, only the first of which has been commented on *ad nauseam*. One dispute, as in a puppet show, pits the wise sage against the blond brute, and is so beautifully staged that the little kids scream in terror that Might will beat down Right. (As we saw earlier, it makes no difference at all if the plot is reworked later by a Nietzschean scriptwriter and now pits the beautiful and sunny Callicles, head of the race of masters, against the black Socrates, degenerate scion of a race of priests and men of *ressentiment*. We, the kids, are still supposed to scream, this time that Right will beat down Might and turn it into a weak and meek sheep.)

But there is a *second* fight going on silently, offstage, pitting the people of Athens, the ten thousand fools, against Socrates and Callicles, allied buddies, who *agree on everything* and differ *only* about the fastest way to silence the crowd. How can we best reverse the balance of force, close the mouths of the multitude, put an end to the disorderly democracy? Will it be through the appeal to reason, geometry, proportion? Or will it be through aristocratic virtue and upbringing? Socrates and Callicles are alone against the crowd, and each of them wants to dominate the mob and obtain a disproportionate share of either this world's or the other world's laurels.

The fight of Might and Right is rigged like a game of catch, and hides the *settlement* between Callicles and Socrates, each agreeing to serve as the other's foil. In order to avoid the fall into Might, let us accept unconditionally the rule of Reason—such has been the earlier version. The later version is the same in reverse: in order to avoid falling into Reason, let us unconditionally agree to fall into the arms of Might. But in the meantime, silent and mute, puzzled and flabbergasted, the people of Athens remain offstage, waiting for their masters to sort out the best way to reverse their "physical force," which could be "entirely discounted" if there were not so many of them. Yes, there are too many, too many to be taken in anymore by this childish story of the cosmic dispute between Might and Right. The hands of the puppeteers are too visible now, and the scandal of seeing Socrates and Callicles, the arch-rivals, arm in arm, is an experience as enlightening for the little kids as seeing the actors of *Hamlet* drink together laughingly at the pub after the curtain has fallen.

Such an experience should leave us older and wiser. Instead of a dramatic opposition between force and reason, we will have to consider *three* different kinds of forces (or three different kinds of reasons—the choice of words adding, from now on, *no decisive nuance*): the force of Socrates, the force of Callicles, and the force of the people. It is a trilogue we have to deal with, and no longer a dialogue. The absolute contradiction between the two famous protagonists is now displaced into a more open contest between two tugs of war: one between the two heroes and the other, not yet recognized by philosophers, between the two heroes pulling on the *same* side of the rope and the ten thousand average citizens pulling on the other side. The principle of the excluded middle that seems so strong in the burning choice between Might and Right—"choose your camp fast or all hell will break loose!"—is now interrupted by *a third party*, the assembled people of Athens. *The excluded middle is the Third Estate.* It sounds better in French: *le tiers exclu c'est le Tiers Etat!* The philosopher does not escape from the Cave, he sends the whole demos down into the Cave to feed only on shadows!

When we hear about the danger of mob rule, we will now be able to ask quietly: "Is it Callicles' solitary rule that you mean, or that of the voiceless assembly of 'human debris and assorted slaves'?" When we hear the little red-flag word "social," we will be able to disentangle two different meanings: the one that designates the power of Callicles' Might against Socrates' Reason, and the one that designates the never-yet-described crowd resisting the attempts of both Socrates *and* Callicles to exert a solitary form of power over them. Two weak, naked, and arrogant men on the one hand; the City of Athens on the other, children, women, and slaves included. The war of two against all, the strange war of the duo trying to make us believe that *without them* it would be the war of all against all.

A Politics Freed from Science
The Body Cosmopolitic

Napoleon's mother used to sneer at her emperor son's fits of rage: "Commediante! Tragediante!" We could mock in the same way these two races of masters, the one descended from Socrates, the other from Callicles. On the comedy side we have the fight between Might and Right; on the tragedy side we have the absolute distinction between *epistèmè* and *pistis,* this *coup de force* whose origin is cleansed by the blood of one martyr. But we can also turn our eyes to the Third Estate and extract from the *Gorgias* the trace of another voice, which is neither comedy nor tragedy but plain prose. Plato is close enough to the benighted time when politics was respected for what it was, that is, before the advent of the scenography set up in common by Socrates and Callicles, which I have defined as "inhumanity against inhumanity." Much as an archaeologist would do with the Delphic Tolos or the statue of Glaucus unearthed by Rousseau, we can thus reconstruct out of the ruins of the dialogue the original Body Politic before it was smashed to pieces—except that I will use the same myth as Rousseau for exactly the opposite goal, that is, to free politics from an excess of reason.

Here is Rousseau in the foreword to the *Discourse on the Origin of Inequality:* "The human soul, like the statue of Glaucus which time, the sea and storms had so much disfigured that it resembled a wild beast more than a god . . . by now we perceive in it, instead of a being always acting from certain and invariable principles, instead of that heavenly and majestic simplicity which its author had impressed upon it, nothing but the shocking contrast of passion that thinks it reasons, and an understanding grown delirious."[1]

By unwinding the adventures of Reason, we can imagine how it was before it turned into an unlivable chimera, a monstrous Big Animal whose unrest horrifies the masters even today. Needless to say, this is an attempt at an archaeology-fiction: the invention of a mythical time when political truth-saying would have been fully understood, a world that was later lost through the accumulation of mistakes and degeneration.

How Socrates Reveals the Virtue of Political Enunciation

In Chapter 7 we noticed many of the specifications of political debate. To reconstruct the virtual image of the original Body Politic, we simply have to take *positively* the long list of negative remarks made by Plato: they show in reverse what is missed when one converts what was, until then, the distributed knowledge of the whole about the whole into an expert knowledge held by a few. Through this bit of archaeology-fiction we can thus be privileged witnesses to two phenomena at once: the specification of the conditions of felicity proper to politics, *and* their systematic destruction by Plato, who turns them into ruins. We thus witness at once the iconoclastic gesture that destroys our much-treasured ability to deal with one another and the conditions of its possible reconstruction.

The dialogue is very explicit about this iconoclasm, since Socrates naively confesses: "In my opinion, you see, rhetoric is a *phantom* of a branch of statesmanship [*politikès morious eidôlon*]" (463d). That is exactly what he and his buddies have done: they have turned a fleshy, rosy living Body Politic that kicked and bit into "a phantom," by asking it to feed on a diet of expert knowledge on which no such organism could survive. They have turned it into an *eidôlon* without realizing that by smashing it they deprived us of one part of our humanity.

1. Rousseau, *Discourse on the Origin of Inequality,* trans. Lester G. Crocker (New York: Pocket Books, 1967).

As Gorgias rightly points out, the first specification of political speech is that it is public and does not take place in the silent isolation of the study or of the laboratory:

> GORGIAS: When I say there's nothing better, Socrates, that is no more than the truth. It [rhetoric] is responsible for *personal freedom* and enables an individual to gain political power in his community.
> SOCRATES: Yes, but what *is* it?
> GORGIAS: I'm talking about the ability to use the *spoken word* to persuade—to persuade the jurors in the courts, the members of the *Council*, the citizens *attending the Assembly*—in short, to *win over* any and every form of public *meeting* of the *citizen body*. (452d–e)

As we just saw, this very specific condition of speaking to all the different forms of assemblies essential to Athenian life (courts, councils, assemblies, burials, ceremonies: all sorts of private and public meetings), is denied by Socrates and turned into a defect, whereas Socrates' weakness, his inability to live in the agora—although he spends all his time in it and seems to enjoy himself immensely!—is vaunted as his highest quality:

> *I'm no politician*, Polus. In fact, last year I was on the Council, thanks to the lottery, and when it was the turn of my tribe to *form the executive committee* and I had to put an issue to the vote, *I made a fool of myself by not knowing the procedure* for this. So please don't tell me to ask the present company to vote now either . . . My expertise is restricted to producing just *a single witness* in support of my ideas—the person with whom I'm carrying on the discussion—and I *pay no attention to large numbers* of people; I only know how to ask for a single person's vote, and *I can't even begin to address people in large groups*. (473e–474a)

Tough luck, because "addressing large numbers" and "paying attention" to what they mean, think, and desire is exactly what is being debated under the despised label "rhetoric." If Socrates is so proud of "not being a politician," why is he teaching those who know better, and why does he not remain in the confines of his own selfish, specialized, expert discipline? What business do agoraphobics have in the agora? This is what Callicles (the real Callicles, the historical, anthropological one whose negative presence can still be detected in the dialogue) rightly points out:

In actual fact, philosophers *don't understand their community's legal system*, or how *to address* either political or private *meetings*, or what kinds of things people *enjoy and desire*. In short, they are completely *out of touch with human nature*. When they do turn to *practical activity*, then, in either a private or a political capacity, they make *ridiculous fools of themselves*—just as, I imagine, politicians make fools of themselves when they're faced with your lot's discussions and ideas. (484d–e)

But Callicles' derision, although it accurately underlines the qualities required from a leader, is itself made useless by his own appeal to an expert knowledge of rhetoric that is content to know nothing at all, to just be manipulative. Yet when he defines the goal of his aristocratic friends, he paints an accurate portrait of the real qualities that Socrates entirely lacks: "The *superior* people I mean aren't shoemakers or cooks: above all, I'm thinking of people who've applied their *cleverness to politics* and thought about how to run their community *well*. But cleverness is only part of it; they also have *courage*, which enables them *to see their policies through to the finish without losing their nerve* and giving up" (491a–b).

It is precisely this courage to see "through to the finish," that Socrates will misrepresent so unfairly when he destroys the subtle mechanism of representation by polluting it with the question of an absolute morality. To see a political project through, with the crowd, for the crowd, in spite of the crowd, is so stunningly difficult that Socrates flees from it. But instead of conceding defeat and acknowledging the specificity of politics, he destroys the means of practicing it, in a sort of scorched-earth policy the blackened wreckage of which is still visible today. And the torch that set the public buildings ablaze is said to be that of Reason!

The second specification that can be recovered from the wreckage is that political reason cannot possibly be the object of professional knowledge. Here the ruins have been so deformed by Plato's iconoclastic obstinacy that they have been made as barely recognizable as those of Carthage. And yet this is what most of the dialogue turns around, as all the commentators have noticed: the question, it appears, is to decide what sort of knowledge rhetoric is. At first, though, it seems very clear that politics is *not* about professionals telling the people what to do: Gorgias says, "I assume you're aware that it was either Themistocles or Pericles, *not the professionals,* whose advice led to those dockyards you mentioned, and to Athens' fortifications and the construction of the harbours" (455d–e).

The protagonists agree that what is needed is not knowledge as such but a very specific form of attention to the whole Body by the whole Body itself. This is what Socrates recognizes under the name of a good and ordered *cosmos* in the qualities required of the expert technicians *(demiourgos):* "Each of them *organizes* the various components he works with into a particular structure and makes them *accommodate and fit one another* until he's formed the whole into an *organized and ordered object"* (503e–504a).

But then, as usual, every time a condition of felicity is clearly articulated it is perverted into its opposite by Socrates, who, as Nietzsche remarked, has King Midas's hands except that he turns gold into mud. The nonprofessional nature of the knowledge of the people by the people turning the whole into an ordered cosmos and not "a disorderly shambles" becomes, through a subtle shift, the *right* of a few rhetoricians to *win over real* experts even if they know *nothing.* What the Sophists meant was that no expert can win in the public agora because of the specific conditions of felicity that reign there. After Socrates' translation, this sensible argument becomes the following absurd one: *any expert* will be defeated by an ignorant person who knows *only* rhetoric. And of course, as usual, the Sophists kindly oblige Socrates by saying the ridiculous thing they have long been accused of saying—this is the great advantage of the dialogue form that *epideixis* lacks:

> SOCRATES: Now, you claimed a while back [456b] that a rhetorician would be *more persuasive* than a doctor even when the issue was health.
> GORGIAS: Yes, I did, as long as *he's speaking in front of a crowd.*
> SOCRATES: By "in front of a crowd" you mean *"in front of non-experts,"* don't you? I mean, a rhetorician wouldn't be more persuasive than a doctor in front of an audience of experts, of course.
> GORGIAS: True. (459a)

Socrates triumphs. Yet again, Gorgias is insisting on the very problem that still besets us today and that no one has ever been able to solve, certainly not Plato and his *Republic.* Politics is about dealing with a crowd of "non-experts," and this situation cannot possibly be *the same thing* as experts dealing with experts in the inner recesses of their special institutions. So when Plato is making his famous joke about a cook and a physician pleading for votes in front of an assembly of spoiled brats (522), it takes very little talent to twist the story to Socrates' embarrassment. This funny scene works only if the crowd of Athens is made up of spoiled kids. Even putting Socrates' aristocratic scorn aside, nowhere does it state, if the story is read carefully, that it pits a serious expert against a populist flatterer. Rather, it stages a *controversy* between two specialists, the cook and the physician, talking to an assembly of grown men about either *short-term or long-term* strategy, the outcome of which neither of them knows, and through which only one party is going to suffer, namely the demos itself.

Here again Socrates' use of a pleasant story hides the dramatic condition of felicity for what it is to speak in real time, in real life, and in full scale about things that no one knows for sure and that affect everyone. About how to fulfill this pragmatic condition he does not have the slightest suggestion, and yet the only solution that the non-experts had in hand—that is, *listening* in the agora to *both* the short-term cook and the long-term physician before running the *risk* of making a decision together that will have legal consequences—is smashed into pieces. We in Europe, who do not know which beefsteak to eat because of the many controversies we read about every day in our newspapers between cooks and physicians about mad cows infected or not by prions, would give several years of our life to recover the solution that Socrates simply *ignores*.

The third condition of felicity is similarly important and similarly ignored. Not only does political reason deal with important matters, taken up by many people in the harsh conditions of urgency, it also cannot rely on any sort of previous knowledge of cause and consequence. In the following passage, which I discussed earlier, the misunderstanding is already clear:

> Rhetoric is an agent of the kind of persuasion *[peithous demiurgos]* which is designed to produce conviction, but not to educate the people, about matters of right and wrong . . . A rhetorician, then, isn't concerned to educate the people assembled in lawcourts and so on about right and wrong; all he wants to do is to *persuade* them *[peistikos]*. I mean, I shouldn't think it's possible for him to get *so*

many people to understand *[didaxai]* such *important* matters *in such a short time.* (454e–455a)

The "demiurge of persuasion" does exactly what the "didactic" urge cannot: it deals with the very conditions of urgency with which politics is faced. Socrates wants to replace *pistis* with a didacticism that is fit for professors asking students to take exams on things known in advance and rehearsed by training and rote exercises, but that is not fit for the trembling souls who have to decide what is right and wrong on the spot. Socrates recognizes this readily: "I think it's a knack *[empeirian]*," he says of rhetoric, "because it *lacks rational understanding* either of the object of its attention or of the *nature of the things* it dispenses (and so it can't explain the reason *[aitian]* why anything happens), and it's inconceivable to me that *anything irrational involves expertise [egô de technèn ou kalô o an è alogon pragma]*" (465a).

How accurate is this definition of what is being destroyed! It is as if we were seeing at once the venerable statue of politics and the hammer that breaks it into pieces. How moving to see, by returning to the past, how close all these Greeks still were to the positive nature of this democracy that remains their wildest invention. Of course "it does not involve expertise," of course "it lacks rational understanding"; the whole dealing with the whole under the incredibly tough constraints of the agora must decide in the dark and will be led by people as blind as themselves, without the benefit of proof, of hindsight, of foresight, of repetitive experiment, of progressive scaling up. In politics there is never a second chance—only one, this occasion, this *kairos*. There is never any knowledge of cause and consequence. Socrates has a good laugh at the ignorant politicians, but *there is no other way* to do politics, and the invention of an afterworld to solve the whole question is exactly what the Sophists laugh at, and rightly so! Politics imposes this simple and harsh condition of felicity: *hic est Rhodus, hic est saltus.*

Here too, after Gorgias points out the real-life conditions in which the *demos* has to reach a decision through rhetoric—"I repeat that its effect is to persuade people *in the kinds of mass-meetings which happen in law courts* and so on; and I think its province is *right and wrong*" (454b)—Socrates requires from rhetoric something it cannot possibly deliver, a *rational* expertise about right and wrong. What could work efficiently with a *relative* difference between bad and good cannot hold water if an *absolute* foundation is required of it, as Socrates demands: "Do you think . . . that all activity aims at the good, and that the good should not be a means towards anything else, but should be *the goal of every action?* . . . Now, is just anyone *competent* to separate good pleasures from bad ones, or does it always take an expert?" (499e–500a).

And Callicles swallows the hook! "It takes an expert," he responds, a *technicos.* From then on, there is no solution and the Body Politic becomes impossible. If there is one thing that does *not* require an expert, and cannot be taken *out* of the hands of the ten thousand fools, it is deciding what is right and wrong, what is good and bad. But the Third Estate has been turned, by Socrates and by Callicles, into a barbaric population of unintelligent, spoiled, and sickly slaves and children, who are now waiting eagerly for their pittance of morality, without which they would have "no understanding" of what to do, what to choose, what to know, what to hope. Yes, "morality is a phantom of statesmanship," its *idol.* And yet, at the same time that Socrates renders the task of politics impossible by asking from the people a knowledge of causes that is totally irrelevant, he defines it accurately: "There's nothing which even a relatively unintelligent person would take *more seriously* than the issue we're discussing—the issue of how to live one's life? The life you're recommending to me involves the *manly activities* of *addressing the assembled people,* rhetorical training, and the kind of *political involvement you and your sort* are engaged in" (500c).

Nothing is more moving in the *Gorgias* than the passage in which Socrates and Callicles, after agreeing on the relevance of statesmanship, destroy, one after another, the only practical means by which a crowd of blind people fumbling in the dark could get the light to help them decide what to do next: "So these are the qualities which that *excellent rhetorical expert of ours* will be aiming for in all his dealings with people's minds, whether he's talking or acting, giving or taking. He'll *constantly be applying his intelligence* to *find ways* for *justice,* self-control, and goodness in all its manifestations to *enter* his fellow citizens' minds, and for injustice, self-indulgence, and badness in all its manifestations to *leave*" (504d–e).

This is what they agree on. This high-minded definition of politics, as we will see, is common sense, but only as long as it is *not* deprived of all the *ways and means* that make it effective. And yet this is what Socrates is going to do, with the straw Callicles following suit obediently.

In a denigration of Athens's beauties that is worse than the city's plunder by the Persians or the Spartans because it comes *from within,* they are going to persuade themselves that every art aims at nothing but corruption. As usual with hearts full of demotic hatred, the loathing for popular culture "blazes forth" every time they talk of politics: "There's *absolutely no expertise* involved in the way it pursues pleasure; it hasn't considered either the *nature* of pleasure or the *reason* why it occurs" (501a).

About what do they talk so irreverently? Cookery first, and then the skills of the greatest playwrights, the greatest sculptors, the greatest musicians, the greatest architects, the greatest orators, the greatest statesmen, the greatest tragedians. All of these people are dumped because they don't know what they know in the didactic fashion that Professor Socrates wants to impose on the people of Athens. Stripped of all its artistic means to express itself to itself, this most sophisticated demos appears this way in the eyes of its disappointed teacher: "So we're faced here with a kind of rhetoric which is addressed to the *assembled population* of men, women, and children all at once—slaves as well as free people—and it's a kind of rhetoric we find *we can't approve of.* I mean, we did describe it as *flattery*"(502d).

Was it simply being flattered to go to the tragedies, to hear the orations, to listen to poetry, to watch the Panathenean's pageantry, to vote with one's own tribe? No, these were the only means by which the demos could accomplish this most extraordinary feat: to represent itself publicly to the public, to render visible what it is and what it wants. All the centuries of arts and literature, all the public spaces—the temples, the Acropolis, the agora—that Socrates is denigrating one by one, were the only ways the Athenians had invented to seize themselves as a totality living together and thinking together. We see here the dramatic double bind that turns the Body Politic into a schizophrenic monster: Socrates appeals to reason and reflection—but then all the arts, all the sites, all the occasions where this reflexivity takes the very specific form of the whole dealing with the whole, are deemed illegitimate. He decries the knowledge of politics for its inability to understand the causes of what it does, but he severs all the feedback loops that would make this knowledge of the cause practical. No wonder Socrates was called the numbfish! What he paralyzes with his electric sting is the very life, the very essence of the Body Politic. How sensible was the Athenian demos to invent this derided institution of ostracism, this very intelligent way to get rid of those who want to get rid of the people!

In this passage the two partners switch off, one by one, each of the hundreds of fragile and tenuous lamps, plunging the *demos* in a darkness much more profound than it was before they started to "enlighten" it—an odious self-annihilation that we cannot mock as a bad show happening on a stage, because it is not Socrates and Callicles who blind themselves; it is we, in the streets, who are deprived of our only fragile lights. No, there is no reason to laugh, because the contempt for politicians is still today what creates the widest consensus in academic circles. And this was written, twenty-five centuries ago, not by a barbaric invader, but by the most sophisticated, enlightened, literate of all writers, who all his life gorged himself on the wealth and beauty that he so foolishly destroys or deems irrelevant for producing political reason and reflection. *This* sort of "deconstruction," not the slow iconoclasm of the present-day sophists, is worth our indignation, because it parades as the highest virtue, and, as Weinberg claims, as our only hope against irrationality. Yes! If there has ever been a form of "higher superstition," it is seen in the dialogue in Socrates' fury for destroying idols and invoking afterworldly, extraterrestrial phantoms.

In a sort of blinding rage, the two sparring partners start killing not only the arts that make reflexivity possible but each of the slightly less blind leaders whose experience was crucially important to the practical politics of Athens: Themistocles, and Pericles himself. This sinister form of iconoclasm does not occur without a concession by Socrates:

> I'm actually *not criticizing them* in their capacity as servants of the state. In fact, I think they were *better* at serving the state than current politicians are . . . However, it's more or less true to say that they were *no better* than current politicians as regards the *only responsibility* a good member of a community has—that is, *altering* the community's needs rather than going along with them, and persuading, or even *forcing*, their fellow citizens to *adopt* a course of action which would result in their becoming *better* people. (517b–c)

But Socrates, as we will see, has deprived the statesmen of all the means to obtain this "alteration," this "betterment," this "forcing function," and so the only thing that is left is either a slavish attachment to what the people think or a mad flight into a fanciful afterworld in which only professors and good pupils would exist. With his inadequate benchmark Socrates takes upon himself the incredible task of passing judgment on all of those who, contrary to what he claims, have led the politics of Athens: "Well, can you name *a single rhetorician* from the past who's supposed to have been instrumental, from his very first public speech onwards, *in changing* the Athenian people from the *terrible state* they'd been in before to a *better* one?" (503b).

To which the only devastating answer is that *no one* has: "It follows from this argument, then, that *Pericles was not a good statesman*" (516d). And the straw Callicles agrees, taking with him the real anthropological Callicles, and Gorgias, and Polus, who of course would have screamed in indignation against this iconoclasm. Instead of defending the great invention of a rhetoric adapted to the subtle conditions of that other great invention, democracy, the straw Callicles shamefully accepts Socrates' judgment.

Among the smoking ruins of those institutions, only one man triumphs: "I'm the *only genuine* practitioner of politics in Athens today, the *only example of a true statesman*" (521d). One man against all! To hide the megalomaniacal dimension of this insane conclusion another folly is added. After mocking rhetoric for providing only "a phantom of statesmanship," Socrates provides an even paler picture. He rules, indeed, but as a shadow, over a demos of shadows: "They'd [the souls] better be judged *naked, stripped* of all this clothing—in other words, they have to be *judged after* they've died. If the assessment is to be *fair*, the judge had better be *naked* as well—which is to say, dead—so that with an *unhampered soul* he can scrutinize the *unhampered soul* of a *freshly dead* individual who isn't *surrounded by his friends and relatives*, and has left those *trappings* behind in the world" (523e).

How right Nietzsche was to put Socrates at the head of his hit list of "men of *ressentiment*." A beautiful scene indeed, this last judgment, but totally irrelevant to politics. Politics is not about "freshly dead" people, but about the living; not about ghoulish stories of the afterworld, but about gory stories of this world. If there is one thing politics does not need, it is yet another afterworld of "unhampered souls." What Socrates does not want to consider is that these attachments, these "friends and relatives," these "trappings," are exactly what obliges us to pass judgment *now*, in the bright sun of Athens, not in the crepuscular light of Hades. What he does not want to realize is that if, by some nightmarish miracle, all of Athens were made of Socrateses who had, like him, shed their wise *pistis* for his didactic knowledge, *none* of the problems of the city would have even begun to be solved. An Athens made of virtuous Socrateses will be no better off if the Body Politic is deprived of its specific form of rationality, this unique circulating virtue, which is like its blood.

How Socrates Misconstrues the Work
Done by the Body Politic upon Itself

Socrates' project is tantamount to replacing the blood of a healthy body with a transfusion from an altogether different species: it can be done, but it is too risky to be done without the informed consent of the patient. If I am using irony and indignation, it is to counterbalance the old habit that makes us either share Socrates' demotic hatred or embrace, without further ado, Callicles' definition of politics as "mere force." What I want to do with this burlesque style is to focus our attention on the middle position, that of the Third Estate which does not ask either for reason or for cynicism. Why is it necessary to make a choice between these two positions, even if this choice paralyzes the Body Politic? As with all choices of this sort, it is because iconoclasm has broken a crucial feature of action (see Chapter 9). An operator that was crucial to the common sense of the common people has been turned into an irrelevant choice—as irrelevant as the epistemologist's unceasing question in Chapter 4, "Are the facts real or are they fabricated?" If we want to speak less polemically, we can say that Socrates' misrepresentation of the Sophists depends on a category mistake. He applies to politics a "context of truth" that pertains to another realm.

The stunning beauty of the *Gorgias* is that this other context is clearly visible in the very lack of comprehension Socrates displays for what it is to *re-present* the people. I am not talking here about the modern notion of representation that will come much later, and that will itself be infused with rationalist definitions, but about a completely ad hoc sort of activity that is neither transcendent nor immanent but more closely resembles a fermentation through which the people brews itself toward a decision—never exactly in accordance with itself, and never led or commanded or directed from above:

"Please tell me, then, which of these two ways of looking after the state you're suggesting I follow. *Is it the one* which is analogous to the practice of medicine and involves *confronting* the Athenian people and *struggling* to ensure their perfection? Or is it the one which is analogous to what servants do and *makes pleasure* the point of the operation? Tell me the truth, Callicles" (521a).

We can ignore for now the childish pleasure Plato takes in making Callicles answer that it is the second, and focus instead on the reason for that choice. The choice is as brutal as it is absurd: either head-on confrontation, the teacher's way, or slavish obsequiousness, the Sophist's way. No teacher, and indeed no servant, has ever behaved like this—and of course no Sophist either. The choice is so bizarre that it can be explained only by Socrates' attempt to bring in a foreign resource, which makes him ask a totally irrelevant question. We know where it comes from. Socrates applies to politics a model of geometrical equality that requires a strict conformity to the model since what is in question is the conservation of proportions through many different relations. Thus the faithfulness of a representation is judged by its ability to transport a proportion through all sorts of transformations. Either it transports it without deformation, and it is deemed accurate, or it transforms it, and it is deemed inaccurate.

As we saw in Chapter 2, in practice the nature of this transformation is precisely to *lose* information on its way and to *re*describe it in a cascade of *re-rep*resentations, or circulating reference, whose precise nature has been as difficult to grasp as that of politics. But thinkers like Plato offered only a theory of how demonstration progressed, not of *its practice.* Thus they could use the idea of a proportion unproblematically maintained through different relations as a benchmark against which to judge all the others. Equipped with this standard, Socrates is going to calibrate every utterance of the poor Sophists: "So that's the course any young member of the community we're imagining must follow if he's wondering how to have a *great deal of power* and avoid being at the receiving end of wrongdoing. He must train himself from an early age to *share the dictator's likes and dislikes,* and he must find a way to *resemble the dictator as closely as he can*" (510d).

Since Socrates voluntarily ignores all the conditions of felicity I listed above, when he evaluates the quality of an utterance it is on the basis of the *resemblance* between the source (here the dictator who represents the spoiled people) and the receptor (here, the young men thirsty for power): "You're so incapable of challenging your loved ones' decisions and assertions *that if anyone were to express surprise at the extraordinary things they cause you to say* once in a while, you'd probably respond—if you were in a truthful mood—by admitting that it's only *when someone stops them voicing these opinions that you'll stop echoing them*" (481e–482a).

Politics is conceived by Socrates as an echo chamber, and there should be no difference between represented and representing except the slight delay that is imposed by the nymph Echo's narrow band width. The same is true for obedience to the master. Once the order is uttered, everyone applies it without deformation or interpretation. No wonder the Body Politic becomes a rather impossible animal: whatever it says, it is always the same thing. Echo for representation, echo for obedience, minus a little bit of static. No invention, no interpretation. Every perturbation is judged a mistake, misrepresentation, misbehavior, betrayal. Imitation for Socrates is necessarily total, either when Callicles repeats what the people say, or when Socrates himself repeats what his true love, philosophy, makes him say (482a), or when statesmen force the people to change their bad ways for better ways (503a). With this benchmark it is easy to see, in Socrates' eyes at least, that Pericles never improved anyone else and that Callicles simply follows the populace: "Now you're terribly clever, of course, but all the same I've had occasion to notice that *you're incapable of objecting to anything your loved ones* say or believe. *You chop and change rather than contradict them.* If in the Assembly the Athenian people *refuse to accept an idea of yours, you change tack and say what they want to hear,* and your behaviour is pretty much the same with that good-looking lad of Pyrilampes'" (481d–e). (Let us remember that in this passage Socrates compares his two loves, Alcibiades and philosophy, with Callicles' two, the Athenian populace and his minion.)

Even here, however, Callicles' behavior—the real Callicles, not the straw one—is perfectly adapted to the ecological conditions of the agora. Far from believing in a "diffusionist" model of information that will travel unadulterated no matter what, he uses an excellent "model of translation" that obliges him to "change tack" when people "refuse to hear his ideas." One can say that Callicles does not hold to truth when he "chops and changes" *only if truth-telling is defined as being convinced alone in the afterworld.* But if the conditions of felicity are, as Callicles so aptly defined them above, for courageous statesmen "to see their policies through to the finish without losing their nerve and giving up," then there is no other way than to negotiate one's position until every one of those who are party to the deal is convinced. In a democracy this means everyone. In the agora there is never any echo, but rumors, condensations, displacements, accumulations, simplifications, detours, transformations—a highly complex chemistry that makes *one* stand for the *whole,* and another chemistry, equally complex, that (sometimes) makes the *whole* obey *one.*

Socrates misjudges the great positive *distance* between what the represented and the representing are saying, because he judges it according to either slavish resemblance or total difference, the only two models he is able to imagine. This is true for representation as well as for obedience. When the citizens repeat what the Body Politic is about, or when they obey the law, none of them slavishly transports without deformation a piece of information. Socrates' dream of replacing all their subtle translations with a strict didactic form of reasoning, like the multiple-choice exams teachers enjoy so much, shows his complete ignorance of what it is to be collectively convinced about matters for which no one has the definite answer. The Sophists in particular had worked out many little tricks and a great treasure of lore to deal with the peculiarity of what cannot be considered an echo chamber or a schoolroom—but their expertise, after Plato's onslaught, is laid waste. The proof is that even here I employ the words "trick" and "lore" to describe an accurate form of knowledge, so powerful is the shadow cast on political reasoning by the notion of information without deformation—the sort of transportation devised as the theoretical justification for geometrical demonstration (see Chapter 2).

Our dialogue catches the specific form of political distance redhanded, so to speak—that is, just when the deed of destruction is being committed. Later, when the iconoclasts have won the day and the dust has settled, the people will be completely unaware that a huge and beautiful statue once stood there. Witness the extraordinary fatherly advice that Socrates gives to Callicles, which accurately defines the proper form of transcendence within which Callicles is still operating and which Socrates is quashing before our very eyes:

If you're under the impression that anyone is going to hand you *the kind of expertise which will enable you to be a political force* here while you're *not assimilated to our system of government* (whether this means that you're better or worse than it), I think you've been *misled*, Callicles. If you're to achieve any *kind of meaningful friendly relationship with the Athenian people* . . . then it's not just *a matter of impersonation: you have to be inherently similar to them*. In other words, it's someone who can *abolish differences between you and them [ostis oun se toutoi omoiotaton apergasetai]* who can turn you into a *rhetorician and the kind of politician you aspire to be,* because everyone enjoys *hearing their own characteristic points of view in a speech and resents hearing anything unfamiliar*—unless you tell me otherwise, my friend. (513a–c)

The real anthropological Callicles would have told him otherwise, if Plato had not held the stylus and turned him into a straw man. "Not only mimesis is sufficient but a complete and total assimilation to the nature of everyone *[ou gar mimètèn dei einai all' autophuôs omoin toutois]*." Never was political reasoning defined so precisely as by the one who rendered it forever impossible. *Autophuôs* says it all, defining with incredible precision that strange form of transcendence and that even stranger kind of reflexivity that remain completely immanent, since, far from the foolish dreams of transparent representation, Socrates endows the Sophists with the power to "grow by themselves" into what all the others are doing and willing. Yes, this is the mysterious quality of politics—which has become a mystery to us but which politicians fortunately preserve with great skill, hidden in their despised tricks and lore.

To read Callicles' calling as immanence, as "assimilation" that "abolishes difference," is to miss the very specific form of transcendence that occurs when the whole represents itself reflexively to the whole, through the mediation of one who takes it upon himself (or herself) to be everyone else—exactly the sort of thing that Socrates is so incapable of doing that he flees from the agora with one or two young men and fulminates against Athens from the safe and nonexistent standpoint of Hades. By reading this alchemy as representation, we miss it exactly as much as Socrates did—and this is the great advantage of the Sophists. They offered a *dark* definition of the Body Politic's "fermentation," instead of the mythically clear self-representation that was invented in the modernist period. Manipulations, differences, combinations, tricks, rhetorics, contribute to that slight difference between the Body and itself. Neither organic bliss nor rationalist transparency: this was the knowledge of the Sophists, expelled from the Republic by the philosopher king.

We are not faced here with one transcendence, Reason, against the immanence of populist leaders, but with *two* transcendences, one admirable to be sure, that of geometrical demonstration, and the other exactly as admirable although utterly distinct, which obliges the whole to deal with itself *without* the benefit of guaranteed information. Viewed from Socrates' remote standpoint, the aim of politics is as impossible as the bootstrapping of Baron von Munchausen. The demos, deprived of knowledge and of morality, needs outside help in order to stand up, and Socrates, generously enough, offers to give it a hand. But if this help were accepted it would not raise the people one inch. The specific transcendence it needs to bootstrap itself is not that of a lever coming from the outside, but much more like the kneading of a dough—except that the demos is at once the flour, the water, the baker, the leavening ferment, and the very act of kneading. Yes, a fermentation, the sort of turmoil that has always seemed so terrible to the powerful, and that has nonetheless always been transcendent enough to make the people move and be represented.

As I said in the previous chapter, the Greeks made one invention too many, either geometry or democracy. But it is a matter of historical contingency that we have inherited this impossible Body Politic. Nothing in principle, except a lack of nerve, obliges us to choose between the two inventions and to forgo part of our rightful heritage. If Socrates had not, by mistake, tried to substitute one type of demonstration, geometry, for another, mass demonstration, *we would be able to honor the scientists without despising the politicians*. It is true that the skills of politics are so difficult, so strenuous, so counterintuitive, and require so much work, so many interruptions, that, to paraphrase Mark Twain's, "there is no extremity to which a man will not go to avoid the hard work of thinking politically." But the mistakes of our forefathers do not prevent us from sorting out their deeds and adopting their good qualities without their defects.

Before we can conclude and restore the two transcendences at once with the fragile plausibility of this archaeology-fiction, we have to understand one last bit about the dialogue. Why is it so often taken as a discussion about morality? I want to argue that, in spite of the moving commentaries by moral philosophers, the ethical questions debated by Socrates and Callicles are so many red herrings. Every time the rhetoricians say something to prove that Socrates' requirements are totally *irrelevant* to the task at hand, Socrates reads it as proof that Sophists are *uninterested* in moral standing. With admirable irony, he delivers, for instance, the following challenge: "Is there anyone—from here or elsewhere, from any walk of life—who was previously bad (that is, unjust, self-indulgent, and thoughtless), but *who has become, thanks to Callicles, a paragon of virtue?*" (515a).

We should not hasten to answer that politics and morality are, of course, two different things, and that, naturally, no one has asked Callicles to turn all citizens into "paragons of virtue"—because if we concede this, we still accept the Machiavellian definition of politics as being *unconcerned* with morality. This would be to live under Callicles' and Socrates' settlement, to take politics as the degraded exercise of conserving power a little longer without any hopes for betterment. This would be playing right into Socrates' hands, because such a disregard for morality is exactly what he wants for the people of Athens without him, and what Machiavelli will later overesteem as a positive definition of political cleverness—although Machiavelli's own position is, of course, not a wholly immoral one.

In fact Plato's perversity is much greater than this. If by morality we mean efforts to provide the Third Estate with ways and means by which to represent themselves to themselves in order to decide what to do next in matters about which there is no definite knowledge, then Socrates is exactly as immoral as Callicles, as I showed earlier, since each is competing with the other over how best to break the majority rule. If anything, Socrates is much worse, since, as we have just witnessed, he systematically destroys what makes representation efficient; whereas Callicles, in spite of Plato's rewrite, still presents, even through his blunders, a vague reminiscence of proper political skills—the real Sophists being dimly visible through their straw counterparts.

Actually, Socrates' crime is mind-boggling because he manages, by a little shift, to take away from the Third Estate exactly the sort of moral behavior that *everyone agrees on,* and then to turn this behavior into an impossible task that can be accomplished only by following his own impossible requirements—the whole thing ending, as we know, in the afterworld of shadows. Quite a feat! And one that, in my view, should be met with grinding of teeth rather than cheers of admiration.

Gorgias, the first to enter the scene, is easily paralyzed by the echo-chamber argument. Exit poor Gorgias. Then Polus is the first to fall into the ethical trap. The question raised by Socrates appears so irrelevant that it works perfectly to divert attention from his own misunderstanding of political representation: "It follows that wrongdoing is the *second worst* thing that can happen; the *worst* thing in the world, the *supreme* curse, is to do wrong and not pay the penalty for it" (479d); "I also claim that to steal, enslave, burgle, and in short to do any kind of wrong against me and my property, is not only *worse* for the wrongdoer than it is for me, the target of his wrongdoing, but is also *more* contemptible" (508e).

We need an enormously long conditioning to see this question as crucially important. Even if morality were taken as nothing more than a sort of basic ethological aptitude of primates in groups, it would still be pretty close to such an assessment. The only thing Socrates adds to turn this into a "big question" is the strict and absolute order of priority that he imposes between suffering wrongdoing and doing it. In exactly the same way as the *absolute* difference between knowledge and know-how was imposed by a *coup de force* for which we had only Socrates' words (see Chapter 7), the absolute difference between what every moral animal believes and what Socrates' higher morality requires is to be imposed by force.

Actually something else is needed, and that is, as usual, the straw Sophists' slavish behavior. It is Polus who makes us believe that we are dealing here with a revolutionary statement: "If you're serious, and if what you're saying really is the truth, surely human life would be *turned upside down*, wouldn't it? Everything we do is the *opposite* of what you imply we *should be* doing" (481c). It is great luck for Socrates that Plato hands him foils like this one, because, without the Sophists' indignation, what Socrates says and what the common people say would be *undistinguishable*. As is usual with revolutionary speeches, there is no safer way to make a revolution than to say that you are making one!

What is so extraordinary is that Socrates, very late in the dialogue, recognizes the obvious commonsense nature of what he has spent so much strenuous effort to prove: "All I'm saying is what I always say: I myself don't know the facts of these matters, but *I've never met anyone, including* the people here today, who *could disagree* with what I'm saying and still avoid making *himself ridiculous*"(509a). Is this not a clear confession that all this long debate with Polus on how to rank moral behavior was never doubted by anyone for any length of time? Everyone is *relatively* bound by the Golden Rule. It is only if you want to turn it into an *absolute* demarcation between suffering and doing evil that it can fail to enlighten you. Exit Polus.

The same paralyzing trick is going to work on poor Callicles who, after appealing, as we saw, to natural laws against conventional laws, is immediately turned into someone who demands unlimited hedonism! This smokescreen is very efficient at hiding how close Socrates' solution is to Callicles' own. And here again, after a lengthy acrimonious *disputatio* in which Callicles conveniently plays the unrestrained beast of prey—as if beasts of prey were themselves unrestrained! as if wolves behaved like wolves, and hyenas like hyenas!—Socrates candidly confesses the basic ethological nature of the morality he, like every slave, child, or, for that matter, chimpanzee (De Waal 1982), relies on: "We shouldn't refuse to restrain our desires, because that *condemns us to a life of endlessly trying* to satisfy them. And this is the life of *a predatory outlaw*, in the sense that *anyone who lives like that will never be on good terms with anyone else*—any other human being, let alone a god— since he's *incapable of co-operation, and co-operation is a prerequisite for friendship*"(507e).

I don't know about the gods, about whom ethological knowledge is slim, but I am confident that even Shirley Strum's baboons and Steve Glickman's hyenas, if they could read Plato, would applaud this description of relative morality in social groups (Strum 1987). The amusing thing is that *no one* ever said the opposite *except* the straw Callicles as portrayed by Plato! The mythology of the war of all against all that threatens to engulf civilization if morality is not enforced is told only by those who have withdrawn from the people the basic morality that sociability has imposed for millions of years on animals in groups. This should be obvious but is not—because, unfortunately, moral philosophy is a narcotic as addictive as epistemology, and we cannot easily kick the habit of thinking that the demos lacks morality as totally as it lacks epistemic knowledge. Even Socrates' admission that what he says is common sense and in no way revolutionary is not enough. Even Callicles' sneering remark that questions of morality are totally irrelevant to the discussion of political rhetoric is not enough: "What I've been thinking about is the *adolescent delight* you take in seizing on any concession someone makes to you, even if he means it as a joke. Do you really think that *I or anyone else would deny* that there are better and worse pleasures?" (499b).

No one denies what Socrates says! No matter what the evidence, moral philosophers portray the *Gorgias* as the magnificent fight of a high-minded Socrates offering the people a goal too high for them to achieve. It is a fight, yes, but one fought by Socrates to impose on the people a definition of morality that they always possessed, *minus* the ways to apply it (Nussbaum 1994)! What Socrates does to the demos of Athens is something as blatantly absurd as if a psychologist, let's say from America, went to China, and working under the very chauvinist conceit that "Chinese people all look alike," decided to paint big numbers on them so as to make them recognizable at last. With what glares will he be met when he arrives with his brush and his bucket of paint and his candid psychological explanation? Can we think that the inhabitants of the huge city of Shanghai will welcome this new way of recognizing one another, because for centuries they have been unable to tell one another apart? Of course not, they will jeer the psychologist away and rightly so, and "his head will spin and his mouth will gape there in that world"! Yet Socrates' use of the morality question in the *Gorgias* is based on exactly the same sort of vast misunderstanding. The Chinese *do* recognize one another without the use of big painted numbers. The demos is endowed with all the morality and all the reflexive knowledge it needs in order to behave itself.

Conclusion: Socrates' Deal and Death

If we bring together all the successive moves that Plato makes Socrates play on the stage, we have a very tricky juggling act:

In the first move, Socrates *takes away* from the people of Athens their basic sociality, their basic morality, their basic know-how, which no one before had ever denied they possessed.

Then, in a second move, stripped of all their qualities, the people are portrayed as children, as beasts of prey, as spoiled slaves ready to attack one another at their slightest whim. Sent down to the Cave, grasping only at shadows, they begin a war of all against all.

Third move: something needs to be done to keep this horrifying mob at bay and set up order against their disorder.

It is at this point that, with trumpet flourishes, the solutions arrive, Reason and Morality. That is the fourth move. But when these are handed back to the people by Socrates, from the exotic realm of geometrical demonstration, the people cannot recognize what has been taken from them, because there is one thing added and one thing missing! What has been added during the passage in the realm of shadows is an absolute requirement that renders morality and know-how inefficient. What has been subtracted is all the practical mediations through which the people could turn their relative knowledge and relative morality to good use in the specific conditions of the agora.

Fifth move: Professor Socrates writes on the blackboard his triumphant equation: politics *plus* absolute morality *minus* practical means *equals* the Impossible Body Politic.

Sixth move, the most dramatic: since the Body Politic is impossible, let us send the whole thing to hell! The *deus ex machina* descends and the three judges of Hades condemn everyone to death—*except* Socrates and "a few other souls"![2] Clap, clap, clap . . .

Let me be naughty (just one last time, I promise) and explain the seventh move, which is the epilogue of this show, which will take place once the crowd has gone home. Is there another explanation, in the end, for this very famous and fair trial, through which the people of Athens forced Socrates to poison himself? To be sure, it was a political mistake, because it made a martyr out of a mad scientist—but it might have been, at least, a healthy reaction against Socrates' most unfair trial of the demos. Was it not fair for someone who wanted to judge naked shadows from the superior seat of eternal justice to be sent to the Isles of the Blessed by the living and fully clothed citizens of Athens? But as we shall now see, this tragicomedy had a great advantage over the later ones: that only one character shed his blood, and he was not part of the public.

2. "Occasionally, however, [Rhadmanthys] comes across a different kind of soul, one which has lived a life of moral integrity, and which belonged to a man who played *no part in public life* or . . . to a philosopher who minded his *own* business and remained *detached from things* throughout his life" (526b–c).

Science Wars? What about Peace?

Let's now abandon the irony and the rage that were needed to press away the poison and extract the honey. We can now salvage from the *Gorgias* the powerful definition of real politics to which epistemic knowledge and absolute morality are so obviously irrelevant. The category mistake is now clear enough. Socrates' and Callicles' settlement can no longer prevent us from liking scientists *as much* as politicians. Contrary to what Weinberg asserts after Plato, there are many possible settlements other than the one I described as "inhumanity to quash inhumanity." A slight change in our definition of science and in our definition of politics will be sufficient, at the end of this chapter, to show the many ways we can now go.

A Science Freed from the Politics of Doing Away with Politics

First let's see, briefly, how the sciences can be freed from the burden of making a type of politics that shortcuts politics. If we now calmly read the *Gorgias,* we recognize that a certain specialized form of reasoning, *epistèmè,* was kidnapped for a political purpose it could not possibly fulfill. This has resulted in bad politics but in an even worse science. If we let the kidnapped sciences escape, then two different meanings of the adjective "scientific" become distinguishable again after being lumped together for so long.

The first meaning is that of Science with a capital S, the ideal of the transportation of information without discussion or deformation. This Science, capital S, is *not* a description of what scientists do. To use an old term, it is an ideology that never had any other use, in the epistemologists' hands, than to offer a *substitute* for public discussion. It has always been a political weapon to do away with the constraints of politics. From the beginning, as we saw in the dialogue, it was tailored for this end alone, and it has never stopped, through the ages, being used in this way.

Because it was intended as a weapon, this conception of Science, the one Weinberg clings to so forcefully, is usable neither to "make humanity less irrational" nor to make the sciences better. It has only one use: "Keep your mouth shut!"—the "you" designating, interestingly enough, other scientists involved in controversies as much as the people in general. "Substitute Science, capital S, for political irrationality" is only a war cry. In that sense, and that sense only, it is useful, as we can witness in these days of the Science Wars. However, this definition of Science No. 1, I am afraid, has no more use than the Maginot Line, and I take great pleasure in being branded as "antiscientific" *if* "scientific" has *only* this first meaning.

But "scientific" has one other meaning, which is much more interesting and is not engaged in doing away with politics, *not* because it is apolitical or because it is politicized, but because it deals with entirely different questions, a difference that is never respected when Science No. 1 is taken, by its friends *as well* as by its foes, as all there is to say about science.

The second meaning of the adjective "scientific" is the gaining of access, through experiments and calculations, to entities that at first do not have the same characteristics as humans do. This definition may seem odd, but it is what is alluded to by Weinberg's own interest in "impersonal laws." Science No. 2 deals with nonhumans, which in the beginning are foreign to social life, and which are slowly socialized in our midst through the channels of laboratories, expeditions, institutions, and so on, as recent historians of science have so often described. What working scientists want to be sure of is that they *do not make up*, with their own repertoire of actions, the new entities to which they have access. They want each new nonhuman to enrich their repertoire of actions, their ontology. Pasteur, for example, does not "construct" his microbes; rather his microbes, and French society, are changed, through their common agency, from a collective made up of, say, *x* entities into one made up of many *more* entities, including microbes.

The definition of Science No. 2 thus alludes to the maximum possible *distance* between standpoints *as different* as possible and to their *intimate* integration into the daily life and thoughts of as many humans as possible. To do justice to this scientific work, Science No. 1 is totally inadequate, because what Science No. 2 needs, contrary to Science No. 1, is lots of controversies, puzzles, risk-taking, imagination, and a "vascularization" with the rest of the collective as rich and as complex as possible. Naturally, these many points of contact between humans and nonhumans are unthinkable either if by "social" we mean Callicles' pure brutal force, or if by "reason" we mean the mouth-shutting Science No. 1. We recognize here, by the way, the two enemy camps between which science studies is trying to gain a foothold: those from the humanities who think we give too much to the nonhumans; and those from some quarters of the "hard" sciences who accuse us of giving too much to the humans. This symmetrical accusation triangulates with great precision the place where we in science studies stand: we follow scientists in their daily scientific practice in the No. 2 definition, not in the No. 1, politicized definition. Reason—meaning Science No. 1—does not describe science better than cynicism describes politics.[3]

So freeing science from politics is easy—not, as has been done in the past, by trying to *isolate* as much as possible the autonomous core of science from the deleterious pollution by the social—but by liberating Science No. 2 as much as possible from the political disciplining that went with Science No. 1 and that Socrates introduced into philosophy. The first solution, inhumanity against inhumanity, relied too much on a fanciful definition of the social—the mob that has to be silenced and disciplined—and on an even more fanciful definition of Science No. 1, conceived as a type of demonstration with no other goal than to bring in the "impersonal laws" to stop controversies from boiling over. The second solution is the best and fastest way to free science from politics. Let Science No. 2 be represented publicly in all its beautiful originality, that is, as what establishes new, unpredictable connections between humans and nonhumans, thus profoundly modifying what the collective is made of. Who defined it most clearly? Socrates—and here I want return to the passage I started with and make amends for having ironized so much at the expense of this master of irony: "In fact, Callicles, the experts' opinion is that co-operation, love, order, discipline, and justice *bind heaven and earth, gods and men.* That's why they

---

3. A third meaning of "scientific" could be added, which I will call *logistics* because it is directly connected to the *number* of entities one wants to access and to socialize. Just as there is a logistical problem to be solved if 20,000 fans are simultaneously trying to park near a baseball stadium, there is a logistical problem to be solved if masses of data have to be transported over a great distance, treated, sorted, "parked," summarized, and expressed. Much of the common usage of the adjective "scientific" refers to this logistical question. But it should not be confused with the other two, especially not with science as access to nonhumans. Science No. 3 ensures that fast and safe communications of data are established; it does not ensure that something sensible is carried over. "Garbage in, garbage out" as the computer motto goes.

call the universe an ordered whole, my friend, rather than a disorderly mess or an unruly shambles [kai to olon touto dia tauta kosmon kalousin, ô etaire, ouk akosmian oude akolasian]" (507e–508a).

Far from taking us away from the agora, Science No. 2—once clearly separated from the impossible agenda of Science, capital S—redefines political order as that which brings together stars, prions, cows, heavens, and people, the task being to turn this collective into a "cosmos" instead of an "unruly shambles." For scientists such an endeavor seems much more lively, much more interesting, much more adapted to their skill and genius, than the boring repetitive chore of beating the poor undisciplined demos with the big stick of "impersonal laws." This new settlement is not the one Socrates and Callicles agreed on— "appealing to one form of inhumanity to avoid inhumane social behavior"—but something that could be defined as "collectively making sure that the collective formed by ever vaster numbers of humans and nonhumans becomes a cosmos."

For this other possible task, however, we not only need scientists who will abandon the older privileges of Science No. 1 and at last take up a science (No. 2) freed from politics, we also need a symmetrical transformation of politics. I confess that this is much more difficult, because, in practice, very few scientists are happy in the artificial straitjacket that Socrates' position imposes on them, and they would be very happy to deal with what they are good at, Science No. 2. But what about politics? To convince Socrates is one thing, but what about Callicles? To free science from politics is easy, but how can we free politics from science?

Freeing Politics from a Power/Knowledge that Makes Politics Impossible

The paradox that is always lost on those who accuse science studies of politicizing science is that it does exactly the reverse, but that, in doing so, it meets another, much stronger opposition than that of epistemologists or of a few disgruntled scientists. If the battle-lines of the so-called Science Wars were drawn in any plausible way, the people like us who are said to "fight" science would be heartily supported by battalions from the social sciences or the humanities. And yet here too it is exactly the reverse. Science No. 2 is a scandal to sociologists and humanists alike because it totally subverts the definition of the social they work with—whereas it is common sense to the scientists, who are worried of course, but only at seeing their unwieldy Science No. 1 taken away from them. The opposition from those who believe in the "social" is a lot more acrimonious than our (on the whole) friendly exchanges with our contradictors from the scientific ranks. How is this possible?

In this too the settlement between Socrates and Callicles can enlighten us, although this is much harder to comprehend. As we saw earlier, when deciphering the tug-of-war between Reason and Force on the one hand and the demos on the other, there are two meanings of the word "social." The first, Social No. 1, is used by Socrates against Callicles (and accepted by the straw Callicles as a good definition of force); the other, Social No. 2, should be used to describe the specific conditions of felicity for the people representing itself, conditions that the *Gorgias* reveals so well even as Socrates smashes them to pieces.

I want to indicate here, as I did in Chapter 3, that the two meanings of "social" are as different as Science No. 1 and Science No. 2. No wonder: the ordinary notion of the social is patterned on the same rationalist argument as that of Science with a capital S—it is a transportation without deformation of inflexible laws. It is called "power" instead of "epistèmè," but this makes no difference, because while epistemologists speak of the "power of demonstrations" sociologists are happy to use their most famous recent motto: "Knowledge/ Power." The damning irony of the social sciences is that when they use this Foucaldian expression to exert their critical skill they in effect say, without realizing it: "Let the agreement of Socrates (Knowledge) and Callicles (Power) stand and triumph *over* the Third Estate!" No critical slogan is less critical than this one, no popular flag is more elitist. What makes this argument difficult to grasp is that natural and social scientists are both behaving as if Power were made of another matter altogether than Reason—hence the supposed originality of separating them and then reuniting them with a mysterious slash. The critics are taken in by Socrates' and Callicles' show. Power and Reason are one and the same, and the Body Politic built by one or by the other is shaped with the same clay; hence the uselessness of the slash, which heightens the interest for the players, and for the critics in their box seats, while boring the audience to tears.

It seems that after the *Gorgias* political philosophy never recovered the full right it once possessed to think over its specific conditions of felicity and to build the Body Politic with its own flesh and blood. The factish*, once smashed, can be patched up but never made whole again. Barbara Cassin, to be sure, has shown beautifully how the second Sophistics won against Plato and reestablished rhetoric *over* philosophy. But this millennium of Pyrrhic victories counted for naught once, in the seventeenth century, another treaty again linked Science and Politics into a common settlement—especially after Machiavelli fell into Socrates' trap and defined politics as a cleverness entirely freed from scientific virtue. Hobbes's Leviathan is a rationalist Beast through and through, made of arguments, proofs, cogs, and wheels. It is a Cartesian *animal-machine* which transports power without discussion or deformation.

Again, Hobbes was used as a foil against reason, much as Callicles was used against Socrates, but the common settlement is even clearer in the seventeenth century than twenty centuries earlier: natural laws and indisputable demonstrations now make for a rationally founded politics. The conditions of felicity for the slow creation of a consensus in the harsh conditions of the agora disappeared underground. There is even less genuine politics in Hobbes than in Socrates' appeal from an afterworld. The only difference is that Socrates' Body Politic has been called back from the dead, to become a Leviathan of *this world*, a monster and a half, composed only of "unhampered" individuals half-dead, half-alive, "without trappings, without clothes, without relatives and friends" (523c)—a scenography altogether more ghoulish than the one imagined by Plato.

Things don't get any better when the Body Politic, to escape from Hobbesian cynicism, is given another transfusion of Reason by Rousseau and his descendants. The impossible surgery started by Socrates continues on an even bigger scale: more Reason, more artificial blood, but less and less of this very specific form of circulating fluid that is the essence of the Body Politic, and for which the Sophists had so many good terms and we so few. The Body Politic is now supposed to be transparent to itself, freed from the manipulations, dark secrets, cleverness, tricks of the Sophists. Representation has taken over, but a representation understood in the very terms of Socrates' demonstration. By pretending to clean Glaucus's statue of all its later deformations, Rousseau makes the Body Politic even more monstrous.

Should I go on with the sad story of how to transform a once-healthy Body into an even more unviable and dangerous monster? No, no one wants to hear more horrific stories, all in the name of Reason. Suffice it to say that when a "scientific politics" is finally invented, then even worse monstrosities come hard and fast. Socrates had only threatened to leave the agora alone, and only *his* blood was shed at the end of his strange attempt at rationalizing politics. How innocent it looks to children of our century! Socrates could not have imagined that scientific programs could later be invented to send the *whole* of the demos into the afterworld and to replace political life with the iron laws of one science—and economics at that! The social sciences in most of their instantiations represent the ultimate reconciliation of Socrates with Callicles, since the brute force advocated by the latter has become a matter of demonstration—not through geometrical equality, of course, but through new tools such as statistics. Every single feature of our definition of the "social" now comes from Socrates and Callicles, fused into one.

I have said enough to make clear why Power/Knowledge is not a solution but yet another attempt to paralyze what is left of the Body Politic. To take Callicles' definition of Power and use it to deconstruct Reason and to show that, instead of the demonstration of truths, Reason involves only the demonstration of force, is simply to reverse the twin definitions that have been devised to make politics unthinkable. Nothing has been achieved, nothing analyzed. It is black and white instead of white and black. The strong hand of Callicles simply takes over from the weakening hand of Socrates the rope used in the tug-of-war against the demos, and later, as the slash indicates, Socrates' hand will take over from the tired hand of Callicles! Admirable collaboration, indeed, but not one that will reinforce the Third Estate, the people pulling on the other end of the rope. To sum up the argument once again, there is not a single trait in the definition of Reason that is not shared by the definition of Force. Thus nothing is gained by trying to alternate between the two or to expand one at the expense of the other. Everything will be gained, however, if we turn our attention toward the sites and situations against which the twin resources of Force/Reason have been devised: the agora.

It is often said that twentieth-century people's bodies are intoxicated by sugar, slowly poisoned by a fabulous excess of carbohydrates unfit for organisms that have evolved for eons on a sugar-poor diet. This is a good metaphor for the Body Politic, slowly poisoned by a fabulous excess of Reason. How ill-adapted was the cure of Professor Socrates is now, I hope, clear, but how much worse is that of the physician *qua* physicist Weinberg, who wants to cure the people's supposed irrationality by bringing in even more "impersonal laws" in order to eliminate even more thoroughly the damnable tendency of the mob to discuss and to disobey. The older settlement had great appeal in the past, even the recent past, because it seemed to offer the fastest way to transform the unruly shambles of gods, heaven, and men, into an ordered whole. It seemed to provide an ideal *shortcut*, a fabulous acceleration, as compared with the slow and delicate politics of producing politics through political means, in the way we learned—and then, alas, unlearned—from the Athenian people. But it has now become apparent that instead of adding order this older solution adds disorder as well.

In the story of the dispute between the cook and the physician with which Socrates amused the public so much, there was some plausibility in this idea of kicking out the cook and letting the physician dictate what we should eat and drink. It no longer applies in our "mad cow" times, when neither the cook nor the physician knows what to tell the assembly, which is no longer made up of spoiled brats and "assorted slaves" but of grown-up citizens. There is a Science War, but it is not the one that pits descendants of Socrates against descendants of Callicles in the rerun of that tired old show: it is the one between "unruly shambles" and "cosmos."

How can we mix Science No. 2, which brings an ever greater number of nonhumans into the agora, with Social No. 2, which deals with the very specific conditions of felicity that cannot be content with transporting forces or truth without deformation? I don't know, but I am sure of one thing: no shortcuts are possible, no short-circuits, and no acceleration. Half of our knowledge may be in the hands of scientists, but the other, missing half is alive only in those most despised of all people, the politicians, who are risking their lives and ours in scientifico-political controversies that nowadays make up most of our daily bread. To deal with these controversies, a "double circulation" has to flow effortlessly again in the Body Politic: the one of science (No. 2) freed from politics, and the one of politics freed from science (No. 1). The task of today can be summed up in the following odd sentence: Can we learn to like scientists as much as politicians so that *at last* we can benefit from the Greeks' two inventions, demonstration and democracy?

The Slight Surprise of Action
Facts, Fetishes, Factishes

What a surprise! I seem to have accomplished my task, to have dismantled the old settlement that held sway over us. The hideout of the kidnappers has been exposed and the nonhumans set free—free, that is, from the squalid fate of providing cannon fodder for the political wars against the demos while clothed in the drab uniform of "objects." This was a perverse politics indeed, the one that aimed at erasing its own conditions of felicity and rendering the Body Politic forever impossible.

And yet it is still as if I have achieved nothing. In the previous chapters I multiplied movements that do not follow the straight path of reason. I proposed many terms to map circuitous moves: labyrinth, translation, shifting out, shifting down. I made great use of metaphors of vascularization, transfusion, connection, and entanglement. To be sure, every time I presented an example my description seemed plausible when it followed the complicated detours made by accurate facts, efficient artifacts, virtuous politics. And yet every time I looked, at a crucial moment, for the term that would allow me to jump, in a single bound, over construction and truth, words failed me. This is not just the usual inadequacy of general words for the particulars of experience. It is as if scientific practice, technical practice, and political practice led into entirely different realms than those of theory of science, theory of techniques, theory of politics. Why is it that we cannot readily recover for our ordinary speech what is so tantalizingly offered by practice? Why is it that associations of humans and nonhumans always become, once clarified, rectified, and straightened out, something so utterly different: two opposing sides in a war between subjects and objects?

Something is missing. Something has been escaping us, chapter after chapter: a way of negotiating a peaceful passage between object and subject, a way of ending this battle without escalating the firepower even more. We need a means of bypassing this standoff altogether, a vehicle, a figure of speech that, instead of *breaking* the subtle language of practice with the intimidating choice "Is it real or is it fabricated? You have to choose, you fools!" would provide a different move, a different register for practice. One thing is sure: once theory has made its analytical cut, once the noise of the breaking bones has been heard, it is no longer possible to account for how we know, how we construct, how we live the Good Life. We are left to try and patch back together subjects and objects, words and world, society and nature, mind and matter—those shards that were made to render any reconciliation impossible. How can we recover our freedom of passage? How can we be trained again to make this swift, elegant, efficient "passing shot," as tennis players say? Why should it be so difficult when everywhere it seems so easy, so widespread? It seems so commonsensical when we listen to the lessons of practice, and yet so contradictory, twisted, and obscure when we hear the lectures of theory.

Where is the solution? *At the point of the break itself.* I want to attempt in this chapter to make us aware of the *very act* of smashing practice into bits. Contrary to what the pragmatists believed (and this is, in my view, why their philosophies never took hold in the public mind), the difference between theory and practice is no more a given than the difference between content and context, nature and society. It is a divide that has been *made*. More exactly, it is a unity that has been fractured by the blow of a powerful hammer.

In the settlement pictured in Figure 1.1, there is one box we have not touched yet, and that is the one labeled "God." I am not alluding here to the moderns' pathetic notion of a God-of-beyond—a supplement of soul for those who have no soul—but to God as the name given to a theory of action, mastery, and creation that served as the foundation for the old modernist settlement. We have interrogated facts and artifacts, we have seen how difficult it is to understand them as being mastered or constructed, but we have not yet inquired into mastery and construction themselves. This is what I want to do now, because I know full well that, without this, no matter how good we are at describing the intricacies of practice we will be immediately attacked as iconoclasts who want to destroy science and morality. I, an iconoclast?! Nothing irritates me more than being presented as provocative or even critical. Especially when such an accusation—or, worse, such a compliment—comes from those who have broken all our figures of speech, from all those descendants of Socrates, one of the first iconbreakers in the lengthy genealogy of idol-smashers who have made us modern. The bitter irony is that iconophiles like me are forced to defend ourselves against iconoclasts. How can this be done? By destroying them and taking our revenge, adding more debris to the debris left by critiques? No, by another means. By *suspending* the crushing blow of the hammer.

Let's start, not at the beginning of this long history as we just did with Socrates, but at its very end. We will take as our exemplar a latter-day iconoclast, one of the courageous critics the moderns have sent around the world to extend the reach of reason, who learn the hard way why they should have suspended their critical gesture instead.

Two Meanings of Agnosticism

His name is Jagannath, and he has decided to break the spell of castes and untouchability by revealing to the pariahs that the sacred *saligram,* the powerful stone that protects his high-caste family, is nothing to be afraid of (Ezechiel and Mukherjee 1990). When the pariahs are gathered in the courtyard of his family estate, the well-meaning iconoclast, to the horror of his aunt, seizes the stone and, crossing the forbidden space that separates the Brahmins from the untouchables in the compound that they share, carries the object to be desecrated by the poor slaves. Suddenly, in the middle of the courtyard, in the blazing sun, Jagannath hesitates. It is this very hesitation that I want to use as my starting point:

> Words stuck in his throat. This stone is nothing, but I have set my heart on it and I am reaching it for you: touch it; touch the vulnerable point of my mind; this is the time of evening prayer; touch; the *nandadeepa* is burning still. Those standing behind me [his aunt and

the priest] are pulling me back by the many bonds of obligation. What are you waiting for? What have I brought? Perhaps it is like this: this has become a saligram because I have offered it as stone. If you touch it, then it would be a stone for them. This my importunity becomes a saligram. Because I have given it, because you have touched it, and because they have all witnessed this event, let this stone change into a saligram, in this darkening nightfall. And let the saligram change into a stone. (101)

But the pariahs recoil in horror:

Jagannath tried to soothe them. He said in his everyday tone of a teacher: "This is mere stone. Touch it and you will see. If you don't, you will remain foolish forever."

He did not know what had happened to them, but found the entire group recoiling suddenly. They winced under their wry faces, afraid to stand and afraid to run away. He had desired and languished for this auspicious moment—this moment of the pariahs touching the image of God. He spoke in a voice choking with great rage: "Yes, touch it!"

He advanced towards them. They shrank back. Some monstrous cruelty overtook the man in him. The pariahs looked like disgusting creatures crawling upon their bellies.

He bit his underlip and said in a firm low voice: "Pilla, touch it! Yes, touch it!"

Pilla [an untouchable foreman] stood blinking. Jagannath felt spent and lost. Whatever he had been teaching them all these days had gone to waste. He rattled dreadfully: "Touch, touch, you TOUCH IT!" It was like the sound of some infuriated animal and it came tearing through him. He was sheer violence itself; he was conscious of nothing else. The pariahs found him more menacing than Bhutaraya [the demon-spirit of the local god]. The air was rent with his screams. "Touch, touch, touch." The strain was too much for the pariahs. Mechanically they came forward, just touched what Jagannath was holding out to them, and immediately withdrew.

Exhausted by violence and distress Jagannath pitched aside the saligram. A heaving anguish had come to a grotesque end. Aunt could be human even when she treated the pariahs as untouchables. He had lost his humanity for a moment. The pariahs had been meaningless things to him. He hung his head. He did not know when the pariahs had gone. Darkness had fallen when he came to know that he was all by himself. Disgusted with his own person he began to walk

about. He asked himself: when they touched it, we lost our humanity—they and me, didn't we? And we died. Where is the flaw of it all,
in me or in society? There was no answer. After a long walk he came
home, feeling dazed. (98–102)

Iconoclasm is an essential part of any critique. But what does the
critic's hammer smash? An idol. A fetish. What is a fetish*? Something that is nothing in itself, but simply the blank screen onto which
we have projected, erroneously, our fancies, our labor, our hopes and
passions. It is a "mere stone," as Jagannath tries to convince himself
and the pariahs. The difficulty, of course, lies in explaining how a fetish could be at once everything (the source of all power for the believers), nothing (a simple piece of wood or stone); and a little bit of
something (that which can reverse the origin of action and make one
believe that, through inversion, reification, or objectification, the object is more than the product of one's own hands). Yet somehow the
fetish gains in strength in the hands of the anti-fetishists. The more you
want it to be nothing, the more action springs back from it. Hence the
disquietude of the well-intentioned iconoclast: "This has become a
saligram *because* I have offered it as a stone."

What has the courageous iconoclast broken? I contend that it is not
the fetish that has been destroyed, but instead *a way of arguing and acting that used to make argument and action possible* and that I now want to
recover ("when they touched it, we lost our humanity—they and me;
didn't we? And we died"). This is the most painful aspect of anti-
fetishism: it is always an *accusation.* Some person, or some people, are
accused of being taken in—or worse, of cynically manipulating credulous believers—by someone who is sure of escaping from this illusion and wants to free the others as well: either from naive belief or
from being manipulative. But if anti-fetishism is clearly an *accusation,*
it is not a *description* of what happens with those who believe or are
manipulated.

Actually, as Jagannath's move beautifully illustrates, it is the critical
thinker who *invents* the notion of belief and manipulation and *projects*
this notion upon a situation in which the fetish plays an entirely different role. Neither the aunt nor the priest ever considered the saligram
as anything but a mere stone. Never. By making it into the powerful
object that must be touched by the pariahs, Jagannath transubstanti-

ates the stone into a monstrous thing—and transmutes himself into a cruel god ("more menacing than Bhutaraya")—while the pariahs are transmogrified into "crawling beasts" and mere "things." Contrary to what the critics always imagine, what horrifies the "natives" in the iconoclastic move is not the threatening gesture that would break their idols but the extravagant belief that the iconoclast *imputes* to them. How could the iconoclast demean himself to the point of believing that we, the natives, should believe so naively—or manipulate so cynically, or fool ourselves so stupidly? Are we animals? Are we monsters? Are we mere things? This is the source of their shame, mistakenly read by the critic as the horror these naive believers should feel when faced with the desecrating gesture that exposes, or so the critic believes, the emptiness of their creed.

In reality the hammer strikes *sideways,* landing on something other than what the iconoclast wanted to break. Instead of freeing the pariahs from their abject condition, Jagannath destroys his own humanity, and that of his aunt, along with the humanity of those he believed he was liberating. Somehow humanity *depended on* the undisturbed presence of this "mere stone." Iconoclasm does not break an idol, but destroys a way of arguing and acting that was anathema to the iconoclast. The only one who is projecting his feelings onto the idol is he, the iconoclast with a hammer, not those who by his gesture should be freed from their shackles. The only one who *believes* is he, the fighter of all beliefs. Why? Because he (I use a masculine pronoun, and that serves him right!) believes in the feeling of belief*, a very strange feeling indeed, one that may not exist anywhere but in the iconoclast's mind.

As we saw in Chapter 5, belief, naive belief, is the only way for the iconoclast to enter into contact, violent contact, with the others—exactly as epistemologists had no other way of contrasting Pasteur and Pouchet than to say that the latter believed and the former knew. Belief, however, is not a psychological state, not a way of grasping statements, but a *polemical* mode of relations. It is only when the statue is hit by the violent blow of the iconoclast's hammer that it becomes a potential idol, naively and falsely endowed with powers that it does not possess—the proof being, for the critic, that it now lies in pieces and nothing happens. Nothing but the indignant bewilderment of those who loved the statue, those who were accused of being taken in by its power and who now stand "liberated" from its sway—but as the novel shows, what lies in ruins in the middle of the desecrated family temple is the humanity of the icon-breaker.

Before it was smashed to bits, the idol was something else, not a stone mistaken for a spirit or any such thing. What was it? Can we retrieve a meaning that would bring the broken pieces back together? Can we, like archaeologists, repair the damage inflicted by time, that greatest of all iconoclasts? We can begin by dusting off the broken shards that we use in our language today, forgetting that they were once joined.

"Fetish" and "fact" can be traced back to the same root. The *fact* is that which is fabricated and not fabricated—as I discussed in Chapter 4. But the *fetish* too is that which is fabricated and not fabricated.[1] There is nothing secret about this joint etymology. Everyone says it constantly, explicitly, obsessively: the scientists in their laboratory practice, the adepts of fetishist cults in their rites (Aquino and Barros 1994). But we use these words *after* the hammer has broken them in two: the fetish has become nothing but an empty stone onto which meaning is mistakenly projected; the fact has become an absolute certainty which can be used as a hammer to break away all the delusions of belief.

Now let us try to glue the two broken symbols together again, to restore the four quadrants of our new repertoire (see Figures 9.1 and 9.2). As we saw in Chapter 4, the fact that is used as a solid hammer is also fabricated, in the laboratory, through a long and complex negotiation. Does the addition of its second half, its hidden history, its laboratory setting, weaken the fact? Yes, because it is no longer solid and sturdy like a hammer (bottom left of Figure 9.1). No, because it is now, so to speak, threadlike, more fragile, more complex, richly vascularized (see Chapter 3) and fully able to generate circulating reference, accuracy, and reality (left side of Figure 9.2). It can still be used, but *not* by an iconoclast and *not* to shatter a belief. A somewhat subtler hand is required to seize this quasi-object, and a somewhat different program of action should be implemented with it.

---

1. One of the inventors of the word "fetishism" links it to another etymology: *fatum, fanum, fari* (De Brosses 1760, 15), but all the dictionaries link it to the Portuguese past participle of "fabricate." On the conceptual history of the term see Pietz 1993, Iacono 1992, and the fascinating enquiry in comparative anthropology of Schaffer 1997.

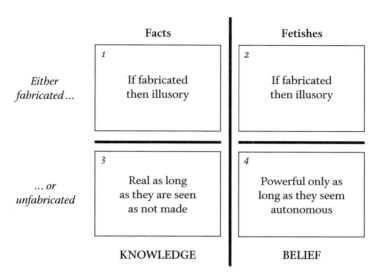

Figure 9.1 In the canonical division of fact and fetish, each of the two divided functions (knowledge and belief) can be exposed by the question: Is it fabricated or is it real? The question implies that fabrication and autonomy are contradictory.

What about the other shard? What happens to the fetish? It is said quite clearly that it has been fabricated, made, invented, devised. None of its practitioners seems to need the belief in belief to account for its efficacy. Everyone is willing to spell out quite frankly how it was made. Does the acknowledgment of this fabrication in any way weaken the claim that the fetish acts independently? Yes, because it is no longer an irresistible ventriloquous phenomenon, an inversion, a reification, an echo, in which the maker is taken in by what it has just created (bottom right of Figure 9.1). No, because it can no longer be seen as a naive belief, as a mere retroprojection of human labor onto an object that is nothing in itself. It is not breakable and fragile like a belief waiting for the iconoclast's hammer. It is now sturdier, much more reflexive, richly invested within a collective practice, reticulated like blood vessels (right side of Figure 9.2). Reality, not belief, is entangled in its filaments. If the hammer's blows threaten it with destruction, they will bounce back from this yielding but resilient network.

If we add to the facts their fabrication in the laboratory, and if we add to the fetishes their explicit and reflexive fabrication by their mak-

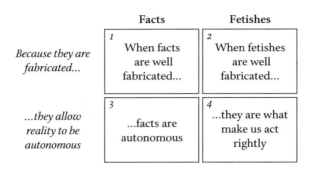

	Facts	Fetishes
Because they are fabricated...	**1** When facts are well fabricated...	**2** When fetishes are well fabricated...
...they allow reality to be autonomous	**3** ...facts are autonomous	**4** ...they are what make us act rightly

FACTISHES

Figure 9.2 Once the fabrication is seen as the cause of autonomy *and* reality for both facts and fetishes, the vertical division between knowledge and belief of Figure 9.1 disappears; it is replaced by a new transversal question: What is it to fabricate *well* so as to make autonomy possible?

ers, the two main resources of the critique disappear: the hammer as well as the anvil (I did not say the hammer and the sickle!). Appearing in their stead is that which had been broken by iconoclasm, and had always been there; that which always has to be carved anew and is necessary for acting and arguing. This is what I call the *factish**. We can retrieve the factish from the massacre of facts and fetishes when we explicitly recover the actions of the makers *of both* (top of Figure 9.2). The symmetry of the two broken symbols is put back into place. If the iconoclast could naively believe that believers exist who are naive enough to endow a stone with spirit (bottom right of Figure 9.1), *it was because the iconoclast also naively believed that the very facts he employed to shatter the idol could exist without the help of any human agency* (bottom left of Figure 9.1). But if human agency is restored in *both* cases (top of Figure 9.2), the belief that was to be shattered disappears, along with the shattering fact. We enter a world that we had never left, except in dreams—the dreams of reason—a world where arguments and actions are everywhere *facilitated, permitted,* and *afforded* by factishes.

The notion of factish is not an analytical category that can be added to others by means of a clear and crisp discourse, since clarity of discourse results from drawing upon the deepest obscurity, being forced to choose between constructivism and reality (the vertical and horizontal axes of Figure 9.1), ushering us to the Procrustean bed in which the modernist settlement wants us all to slumber: Are scientific facts real or are they constructed? Are fetishes beliefs that are projected on idols or are these idols "really" acting? Although these questions are commonsensical enough, and seem necessary for any analytical clarity, they are, on the contrary, the questions that render all of the associations between humans and nonhumans totally opaque. If there is one thing that obscures the saligram's function, it is asking whether or not it is a "mere" stone, a powerful object or a social construction.

But if one refuses to answer the question "Is it real or is it constructed," a serious problem can arise. Answering with the agnostic's "no comment," can easily be confused with a cynical acceptance of the falsity of all human representations. This, as I said at the end of Chapter 1, is where science studies flirts dangerously with its polar opposite, postmodernism. The solution of the factish is not to *ignore* the choice, as many postmoderns do, by saying, "Yes, of course, construction and reality are the same thing; everything is just so much illusion, storytelling, and make believe. Who would be so naive, nowadays, as to dispute such trivia?" The factish suggests an entirely different move: it is *because* it is constructed that it is so very real, so autonomous, so independent of our own hands. As we have seen over and over, attachments do not decrease autonomy, but foster it. Until we understand that the terms "construction" and "autonomous reality" are *synonyms*, we will misconstrue the factish as yet another form of social constructivism rather than seeing it as the modification of the entire theory of *what it means to construct*.

Another way to say this is to point out that the modernists and postmodernists, in all their efforts at critique, have left belief, the untouchable center of their courageous enterprises, untouched. They believe in belief. They believe that people naively believe. There are thus two forms of agnosticism. The first one, so dear to the critics' hearts, consists of a selective refusal to believe *in* the content of belief—usually God; more generally fetishisms and such things as saligrams; more recently popular culture; and eventually scientific facts themselves. In this definition of agnosticism the thing to be avoided at all costs is being taken in. Naïveté is the capital crime. Salvation always comes from revealing the labor that hides behind the *illusio* of autonomy and independence, the strings that hold the puppets up. But I will define *agnosticism*, not as the doubting of values, powers, ideas, truths, distinctions, or constructions, but as the doubt exerted *against this doubt* itself, against the notion that *belief* could in any way be what holds any of these forms of life together. If we do away with belief (in beliefs) then we can explore other models of action and mastery. Before we can do this, however, we have to take a last quick look at the modern critic.

A Sketch of the Modern Critic

There is some difficulty in my speaking as though only the iconoclast is a naive believer, as if he and he alone projects feelings onto objects and forgets that the facts he makes in the laboratory are not the products of his own hands. How could he and he alone be naive, immersed in bad faith, and blinded by false consciousness? Am I not displaying a lack of charity here, or worse, a lack of reflexivity? It is true that the modern iconoclast does not believe more naively in his double construction of facts and fetishes than any of the others believed in the idols that the iconoclast destroyed to "free" them from their chains. Something else is at stake in his obsession, a different wisdom, which, to be sure, is not that of the factish, but is a wisdom all the same, no matter how tortuous it appears. Let us consider one last time the extraordinary power of the modern iconoclast in his native habitat, when he is not being self-conscious, that is, before he stops being modern, when he still possesses his pristine and unspoiled exoticism, in the very moment at which he tries, like Jagannath, to desecrate what he believes to be a mere stone that common people endow with nonexistent powers!

Is the modernist critic imprisoned and enchained by his delusionary and muddled belief? On the contrary; the belief that *others* believe is a very precise mechanism that allows the human an extraordinary degree of freedom. *By removing human agency twice,* it makes it possible, at no cost, to free the passage for action, to clear the path by disintegrating entities into mere beliefs and solidifying opinions and positions into hard facts. No one has ever had so much freedom. Freedom is precisely what permits and justifies the iconoclast's strokes. But freedom from what? *Freedom from caution and care,* as I will discuss in the next section.

We see now that the iconoclast is not free from factishes, because he cannot escape the human agency that manufactures facts in the laboratory; nor is he free to do away with entities by confining them in internal states of a mind endowed with an imagination and a "deep" unconscious. In this regard modernists are like everyone else: everyone everywhere has need of factishes to act and to argue. There is only one nonmodern humanity—and in this sense, yes, I believe in a universalist anthropology. But the main cunning of the critical modernist lies in his ability to use the *two sets* of resources at once: on the one hand the factishes, like everyone else, and on the other the apparently contradictory theory that radically distinguishes facts (which no one has made), from fetishes (which are totally nonexistent objects, simply beliefs and internal representations)—see the two columns of Figure 9.1. This is what makes the modernist a true anthropological curiosity, this is his unique and incommensurable "genius" which allows comparative anthropology to recognize *this* culture among all the others.

How can you recognize a modernist? Let us very quickly list the items on the modernist's psycho-social profile.

Modernists are iconoclasts. They have all the rage and violence and power that allows them to break the factishes and to produce two irreconcilable enemies: fetishes and facts.

Modernists are freed, by this very act of shattering, *from the chains* that bind all other cultures, since they can, at will, pump out of existence whatever entities restrict their action, and pump into existence whatever entities will enhance or accelerate their action (at least this is the way they used to understand the "other cultures," as if these were "blocked," or "limited," or "paralyzed").

Modernists, protected by this iconoclasm, can then proceed like everyone else to produce, inside the insulated wombs of their "laboratories," as many factishes as they want. To them even the sky is not a limit. New hybrids can be launched endlessly since there are no consequences attached to them. The inventiveness, originality, and juvenile ardor of the moderns can flourish unfettered. "This is only practice," they can say, "it has no consequence; theory will remain safe for ever." Modernists behave like the Carthaginians, who say, as they sacrifice their own children to Baal, "They are only calves, only calves, not children!" (Serres 1987).

High above them, watching like protective goddesses, the sharp-cut distinctions between subject and object, science and politics, facts and fetishes, render forever invisible the complicated and rather bizarre means by which all of these categories are mixed. Above, subjects and objects are infinitely distant, especially in theories of science. Below, subjects and objects are intermingled to the greatest extreme, especially in the practice of science. Above, facts and values are kept infinitely far apart. Below, they are confused and redistributed and tossed around endlessly. Above, science and politics never mix. Below, they are continually remade anew from top to bottom.

Notice the construction that makes factishes thrice invisible: above, they have disappeared, replaced by a clear and radiant theory whose blinding light is fueled by a complete and constant distinction between fact and fiction; below, the factishes are there—how could they not be?—but they are hidden, invisible, mute, since only silent and babbling practice* can account for that which is strictly forbidden above. To be sure, actors constantly speak about "that," the vast cauldron at the heart of all their projects, but in a shattered and hesitant language that only fieldwork can retrieve, and that never threatens the opposed discourse of theory. Finally, an absolute distinction keeps the top of the setup separate from the bottom part. Of course the factishes of the modern exist, but their construction is so strange that although they are active everywhere, visible to the naked eye, they remain invisible and impossible to register.

Naturally, however, the moderns are conscious, reflexive, and explicit about this threefold construction. We are not dealing here with a "superego" of theory, obsessively silencing the "id" of practice. If they were not conscious, we would need another conspiracy theory, another psychoanalysis, to account for the belief in belief, to explain the modernists' belief in *illusio,* and to deny to the moderns, and only to them, the right to be like everyone else, that is, to be free from belief, in the firm hands of factishes—and I, for one, would be forced to become the iconoclast who would reveal the harsh reality of practice behind the veil of theory.

How do we know the moderns are aware that they have never been modern? Because, far from keeping the facts separate from the fiction and the theory of this separation from the practice of mediation, they endlessly, obsessively fix up, repair, and overcome these broken fragments. They use everything at hand to show that subjects and objects should be reconciled, patched up, overtaken, "aufhebunged." Modernism never stops repairing, and patching up again, and being desperate about not being able to fix it because, despite all this repair work, modernists never abandon the shattering gesture that started it all, the one that created modernity in the first place. So desperate are they that, after having shattered all the other cultures, they start to *envy* them and to devise, under the name of exoticism, the museographic cult of the whole, complete, organic, wholesome, unspoiled, untouched, unmodernized savage! To the modern they add an even more bizarre invention, the premodern*.

We can now sketch the ideal psychosocial type of the modern, the model of the critique. As an iconoclast, the modern breaks the idols, all of them, always, fiercely. Then, protected by this gesture, in the silent practice opened up for him like a huge underground cavity, he can get his kicks, with all the juvenile enthusiasm of the inventor, from mixing up all sorts of hybrids without fearing any of the consequences. No fear, no past, only more and more combinations to try. But then, terrified by a sudden realization of the consequences—how could a fact be just a fact with no history, no past, and no consequence, a "bald" fact instead of a "hairy" one?—he suddenly shifts from brave iconoclasm and youthful ardor to fits of guilt-ridden bad conscience: and this time he destroys himself, in endless ceremonies of atonement, looking everywhere for the broken fragments of his creative destruction, gathering them back into huge and fragile bundles.

The strangest thing is that these godless, fetishless creatures are viewed by all the others as having terrifying protectors and gods! And the other cultures cannot decide when the moderns are at their most terrifying: is it when they crush the idols and burn them in *autos-da-fé?* Is it when they innovate freely in their laboratories without the slightest worry about the consequences? Or is it when they go around beating their chests and tearing out their hair, desperately inflicting penance on themselves for the sins they have committed, trying to recover in their museums, films, retreats, and self-help books the wholeness of the lost paradise? "The pariahs found him more menacing than Bhutaraya"—which means that the freedom fighter now has the power of *three* gods on his side instead of one: the menacing head of the master Brahmin, the menacing force of modernization, and the

power of the local god. Whether the struggle for modernization succeeds or not, it always seems to be the pariahs who end up losing.

Yes indeed, the moderns are interesting characters, well worth the
attention of comparative anthropologists!

Another Theory of Action and Creation

Now that we have turned the modernist repertoire from a resource to
a topic for study, now that we have portrayed the guilt-ridden iconoclast as one interesting but peculiar type in one culture among others,
is it possible to imagine a model for the practice of politics that would
not rely so heavily on the model of the critique? This is a difficult
question because the scenography of activism has been so powerfully
based on iconoclasm that it seems that if you do away with the iconoclasm you immediately fall into one of a very few models of reactionary politics. If one is neither modern nor premodern, is not the only
alternative left that of being antimodern? How can the number of
models for political action be multiplied; how can we undo the current definitions of "reactionary" versus "enlightened" politics? One
way is to modify the scenography of politics itself, as I attempted to do
 in Chapters 7 and 8. Another path, which I took in Chapter 6, is to offer an alternative to the idea of progress that still makes use of the
traditional arrow of time. The possibility that I want to outline now
requires us to consider what sort of life we would lead if we lived under the protection of factishes again—no longer caught between facts
and fetishes. At least three things would change profoundly: the definition of action and mastery; the dividing line between a physical
world "out there" and a mental world "in there"; and the definitions of
care and caution along with the public institutions that would exhibit
them.

Action and Mastery

What is it that iconoclasm breaks, and what is it that factishes allow us
to restore? A certain theory of action and of mastery. Once the hammer has fallen, shattering the world into facts on one side and fetishes
on the other, nothing can stop the dual question from being posed: did
you construct the thing yourself, or is it autonomous? This ceaseless,
sterile, and boring question paralyzed the field of science studies cen-

turies before it even began. When a fact is fabricated, who is doing the fabrication? The scientist? The thing? If you answer "the thing," then you are an outdated realist. If you answer "the scientist," then you are a constructivist. If you answer "both," then you are doing one of those repair jobs known as the dialectic, which seem to patch up the dichotomy for a while, but only hide it, allowing it to fester at a deeper level by turning it into a contradiction that has to be resolved and overcome. And yet we have to say that it *is* both, obviously, but without the assurance, certainty, or arrogance that seems to go with the realist *or* the relativist answer or with a clever oscillation between the two. Laboratory scientists make autonomous facts. That we have to hesitate between two versions of this simple "making do" *(fait-faire)* proves that we have been hit by a hammer that has broken the simple and straightforward factish into two parts. The shock of critical intelligence has rendered us stupid.

Things change entirely, as we saw in Chapter 4, when we listen to what is said by practicing scientists without adding or withdrawing anything. The scientist makes the fact, but whenever we make something *we* are not in command, we are slightly *overtaken* by the action: every builder knows that. Thus the paradox of constructivism is that it uses a vocabulary of *mastery* that no architect, mason, city planner, or carpenter would ever use. Are we fooled by what we do? Are we controlled, possessed, alienated? No, not always, not quite. That which slightly overtakes us is *also*, because of our agency, because of the *clinamen* of our action, slightly overtaken, modified. Am I simply restating the dialectic? No, there is no object, no subject, no contradiction, no *Aufhebung*, no mastery, no recapitulation, no spirit, no alienation. But there are events*. I never *act;* I am always slightly surprised by what I do. That which acts through me is also surprised by what I do, by the chance to mutate, to change, and to bifurcate, the chance that I and the circumstances surrounding me offer to that which has been invited, recovered, welcomed (Jullien 1995).

Action is not about mastery. It is not a question of a hammer and shards, but one of bifurcations, events, circumstances. These subtleties are difficult to retrieve once iconoclasm has struck, because facts and tools are now firmly in place, suggesting the model for *Homo faber* that can never, after that, be displaced and reworked. But, as we saw in Chapter 6, no human agent has ever built, constructed, or fabricated anything, not even a stone tool, not even a basket, not even a bow, by using the repertoire of action invented for *Homo faber*. *Homo faber* is man's fable, a *Homo fabulosus* through and through, a retrospective projection into our fantastic past of a definition of matter, humanity, mastery, and agency that dates entirely from the modernist period, and that uses only a quarter of its repertoire—the world of inert autonomous matter. We cannot account for laboratory practice by falling back on a modernist definition of technical construction—or, even less, one of social construction.

Why is it so difficult to retrieve other theories of action? Because it is crucially important to the modernist ethos to demand a choice between that which you fabricate—as a free and naked human—and that which is a fact out there, made by no one. The whole work of the modern has been to render these two agents, the human and the object, unfit for any other role than that of opposing each other. No wonder they cannot be used for anything else! It is a simple question of ergonomics: they aren't suited for any other job.

But the idiom changes immediately once the two halves are brought together again. Facts are fabricated; we make facts, that is, there is a *"fait-faire."* Of course the scientist does not make up facts—who has ever made *up* anything? This is another fable, symmetrical with that of *Homo faber* and dealing, this time, with the fancies of the mind. I do not deny that people have minds—but the mind is not a world-creating despot that makes up facts to suit its fancy. Thought is seized, modified, altered, possessed by nonhumans, who, in their turn, given this opportunity by the scientists' work, alter their trajectories, destinies, histories. Only modernists believe that the only choice to be made is between a Sartrean agent and an inert thing out there, a root on which to vomit. Every scientist knows in practice that things have a history too; Newton "happens to" gravity, Pasteur "happens to" the microbes. "Intermingle," "bifurcate," "happen," "coalesce," "negotiate," "ally," "be the circumstances of": these are some of the verbs that signal the shift in attention from the modernist to the nonmodernist idiom.

What is at stake here is mastery. In making the world the product of individuals' thoughts and fancies and in talking about construction as though it involved the free play of fancy, modernists believe they make the world in their image, just as God made them in his. This is a strange and rather impious description of God. As if God were master of His Creation! As if He were omnipotent and omniscient! If He had all these perfections, there would be no Creation. As Whitehead so beautifully proposed, God, too, is slightly overtaken by His Creation, that is, by all that is changed and modified and altered in encountering Him: "All actual entities share with God this characteristic of self-causation. For this reason every actual entity also shares with God the characteristic of transcending all other actual entities, *including God*" (Whitehead [1929] 1978, 223, my italics). Yes, we are indeed made in the image of God, that is, *we* do not know what we are doing either. We are surprised by what we make even when we have, even when we believe we have, complete mastery. Even a software programmer is surprised by her creation after writing two thousand lines of software; should God not be surprised after putting together a much larger package? Who has ever mastered an action? Show me a novelist, a painter, an architect, a cook, who has not, like God, been surprised, overcome, ravished by what she was—what *they* were—no longer doing.

And do not tell me they were "possessed," "alienated," or "dominated" by outside forces. They never exactly say so. They say that these others have been modified, altered, taken over, in the circumstances of the action, by the unfolding of the event. Mastery, domination, or recapitulation is not the way to think of such instances. No nonmodern wants to have to deal with that sort of God or that sort of Man. Factishes bring with them a quite different definition of God, of human agency, of action, of nonhumans. No model of political action can be offered as an alternative to the model of the critique until we modify our anthropology of creation, that is, until we retrieve the anthropology *practiced* by the modernists even while they believed themselves to be modern, and while they always explicitly said, in practice, that they were not.

An Alternative to Beliefs

Is it really possible to be agnostic in the sense I have defined? Is not belief in belief what allows the distinction between a world "out there" and a palace of ideas, imagination, fancies, and distortions "in there"? How could we survive without this distinction between epistemological and ontological questions? Into what sort of obscurantism would we fall if we could no longer make the sharp distinction between the contents of our heads and the world outside our minds? And yet the price paid for obtaining this semblance of common sense is extraordinarily high. We are so used to living under the sway of anti-fetishism, so accustomed to taking for granted the abyss between the wisdom of practice and the lessons of theory, that we seem to have entirely forgotten that this most cherished analytical clarity was reached at the price of an incredibly costly invention: *one physical* world "out there" versus *many mental* worlds "in there." How did this come about?

If, as common sense would have it, there are no factishes, but only fetishes, which are nothing but pieces of wood and mute stones, where can all those things that believers believe in be located? There is no solution but to push them into the *minds* of believers or into their fecund imaginations, or to embed them even deeper, in a rather perverse and crooked unconscious. Why not leave them where they were, that is, among the multiplicity of nonhumans? Because there is no longer any room for nonhumans or for any multiplicity. The world itself has been stuffed beyond capacity, thanks to the *other, simultaneous move* that transformed factishes into facts. If no human agency is at work—or has been at work—in the manufacture of facts, if there are no limits of cost, information, networks, or manpower for the production, expansion, and maintenance of facts, then nothing, absolutely nothing stops them from proliferating everywhere, continuously, filling in every last corner of the world—and at the same time unifying the many worlds into a single homogeneous one. The notions of matter, a mechanical universe, a mechanical world-picture, a natural world: these are the simple consequences of the rupture between the two meanings of "fact"—that which is fabricated, that which is not fabricated. But the notions of belief, mind, interior representation, illusion: these are merely the consequences of having split the factish in two—that which is fabricated, that which is not fabricated.

It is hard to decide which came first. Was the notion of an interior mind invented as a repository for all the entities squeezed out of the world, or did the belief in beliefs empty the world, leaving room for "factoids" to proliferate like rabbits in Australia? What is certain is that with the destruction of the means of argumentation and action that factishes enabled, with the removal of human agency from the fabrication of facts *and* from the fabrication of fetishes, two fabulous reservoirs were invented, *one for epistemology, one for ontology*. These subjects endowed with an inside are as strange as these objects relegated to an outside. Indeed, the notion of an inside divided from an outside is very strange and is, in its own right, a fabulous innovation. With one stroke the iconoclast sets in motion the most powerful suction pump ever devised. Whenever entities are obstacles to his action they can be pumped out of existence, emptied of all reality until they are nothing but hollow beliefs. Whenever there is a deficit of certain, positive mechanical entities to render his actions steady and beyond objection, they can be pumped into existence by the score: now there are stones everywhere "out there," in the only world there is, matched by many naive beliefs about saligrams "in there," inside the believers' minds. With this device, powered by the opposition between epistemology and ontology, the iconoclast is able to empty the world of all its inhabitants by turning them into representations, while filling it up with continuous mechanical matter.

But what happens when this pump has stalled, when there is no longer an inside mind into which, under the name of fancy or belief, one can squeeze every entity, and when there is no longer an outside world made of ahistorical, inhuman causes "out there"? The first thing to go, naturally, is the very difference between inside and outside. This does not mean that everything is now outside, but simply that the entire scenography of outside and inside has evaporated.

What appears in its place is, at first, as we witnessed in Exhibit A in Chapter 5, a bewildering array of entities, divinities, angels, goddesses, golden mountains, bald kings of France, characters, controversies about facts, propositions in all possible phases of existence. The stage will be so crowded with this heterogeneous crew that one may start to worry, and get nostalgic for the good old modern age when the pump was still at work, sucking all beliefs out of existence and replacing them with sure and safe and certain objects of nature. But fortunately these entities do not ask for the same kinds of ontological *specifications*. They cannot be ordered, to be sure, into beliefs and realities, but they can be ordered, and very neatly, according to the types of existence they claim.

Jagannath's stone, for instance, does not claim to be a spirit as in the fetishist mode, nor does it claim to be the symbol for a spirit projected onto the stone, as in the anti-fetishist version. As Jagannath realizes clearly when he fails to desecrate the saligram, this stone is what makes him, his family, and the untouchables human, what holds them in existence, that without which they would die. Understood according to the fact-fetish dichotomy, the stone immediately becomes a spirit, that is, a transcendent entity that obeys the *same* specifications as an object of nature *except* that it is invisible. In practice, however, the stone is a factish and does not claim to be a spirit, to be invisible; it never ceases to be, even for the aunt and the priest, a "mere stone." It simply asks to be that which *protects humans against inhumanity and death*, the thing that, when removed, turns them into monsters, animals, things (Nathan and Stengers 1995).

The problem is that this way of arguing—granting ontological content to beliefs—runs counter to the whole deontology of the social sciences. "When the sage points at the Moon," says the Chinese proverb, "the fool looks at his fingertip." Well, we have all educated ourselves to be fools! This is our deontology. This is what a social scientist learns at school, mocking the unwashed who naively believe in the Moon. *We* know that when actors speak about the Virgin Mary, divinities, saligrams, UFOs, black holes, viruses, genes, sexuality, and so forth we should *not* look at the things thus designated—who could be so naive nowadays?—but should look *instead* at the finger, and from there, following down the arm along the nerve fibers, to the mind of the believer, and from there down the spinal cord to the social structures, the cultural systems, the discursive formations, or the evolutionary bases that make such beliefs possible. The anti-fetishist bias is so strong that it seems impossible to argue against it without hearing the indignant screams: "Realism! Religiosity! Spiritism! Reaction!" We should now imagine a scene that would enact Jagannath's trauma, but in reverse: the nonmodern thinker wants to touch the *contents* of beliefs again, and the modernist and postmodernist critics, horror stricken, scream, "Don't touch them!! Don't touch them! Anathema!" And yet we, the science students, have touched them, and nothing happened except that the dreams of social constructivism disappeared! Through a transfiguration exactly opposed to that of Jagannath, when we touched subjects and objects they suddenly turned into humans and nonhumans.

After centuries of detachment, our attention is now turning back to the fingertip, and from it to the Moon. The simplest explanation for all the attitudes of humanity since the dawn of its existence is probably that people mean what they say, and that, when they designate an object, this object is the cause of their behavior—*not* a delusion to be explained by a mental state. Here again we should understand that the situation has changed entirely since the advent of science studies. It was feasible to be anti-fetishist when facts could be used as destructive weapons against beliefs. But if we now speak of factishes, there exist neither beliefs (to be fostered or destroyed) nor facts (to be used as a hammer). The situation has become much more interesting. We are now faced with many different practical metaphysics, many different practical ontologies.

By granting ontology back to nonhuman entities, we can begin to tackle the major question at issue in the science wars. The modernist Enlightenment, in its republican ideal at least, became, for a while, a popular movement. It struck a chord in all the oppressed around the world. When facts were accommodated into our collective existence, great clouds of delusion, oppression, manipulation were dissipated. But since then the models offered by the critique have ceased to be popular. They now run against the very grain of what it is to be human and to believe. Facts have gone too far, attempting to transform everything else into beliefs. The burden of supporting all these beliefs becomes unbearable when, as in the postmodern predicament, *science itself* is submitted to the same doubt. It is one thing to attack beliefs when we are fortified by the certainties of science. But what are we to do when science itself is transformed into a belief? The only solution is postmodern virtuality—the nadir, the absolute zero of politics, aesthetics, and metaphysics. The engine of virtuality, however, is in postmodern heads, not in the worlds surrounding them. Virtuality is what everything else turns into when belief in belief has run amok. It is time to stop the little salt-mill's grinding, before everything left becomes bitter.

Could we not say, quite simply, that people are tired of being accused of believing in nonexistent things, Allah, jinns, angels, Mary, Gaia, gluons, retroviruses, rock 'n' roll, television, laws, and so on? The nonmodern intellectual does not take Jagannath's position, day after day bringing new saligrams to desecrate and then throwing them aside, discouraged to discover that only he, the desecrator, the iconoclast, the liberator, believes in them and that everyone else—ordinary pariahs, average laboratory scientists—has always lived under a completely different definition of action, in the hands of factishes of totally different shapes and functions.

Care and Caution

What did the factish do before it was broken by the anti-fetishist's blow? To say that it mediated action between construction and autonomy is an understatement, and relies too heavily on the ambiguity of the term mediation*. Action is not what people do, but is instead the *"fait-faire,"* the making-do, accomplished along with others in an event, with the specific opportunities provided by the circumstances. These others are not ideas, or things, but nonhuman entities, or, as I called them in Chapter 4, propositions*, which have their own ontological specifications and populate, along their complex gradients, a world that is neither the mental world of psychologists nor the physical world of epistemologists, although it is as strange as the first and as real as the second.

What the factishes were good at was articulating *caution* and *publicity*. They publicly declared that care should be taken in the manipulation of hybrids. When they tried to break the fetishes, the iconoclasts broke the factishes instead. As I have said, these rampages are what have given the moderns their fabulous energy, invention, and creativity. They are no longer held by any constraints, any responsibility. The broken halves of the factish, nailed above the threshold of the modernist temple, protect them against all the moral implications of what they do, and they can be all the more inventive since they believe themselves to be wallowing in "mere practice." What has been removed by the hammer is care and caution.

Of course, action did have consequences, but these came later, literally *after the fact*, and under the subservient guise of unexpected consequences, of belated impact (Beck 1995). Modernist objects were bald—aesthetically, morally, epistemologically—but the ones produced by the nonmoderns have always been hairy, networky, rhizomelike. The reason one should always beware of factishes is that their consequences are unforeseen, the moral order fragile, the social one unstable. This is just what modernist facts have shown us over and over, except that, for the modern, consequences are nothing but an afterthought. It is only *after* the desecrating ceremony that Jagannath realizes that no one ever believed the saligram to be anything but a stone, and that the only inhumanity was that which he, the free-thinker, produced by destroying the idol. When the aunt and the priest screamed "Beware! Beware!" they did not mean, *as he thought,* that they were afraid he would break the taboo, but that they were afraid he would break the factish that kept care and caution under attentive public consideration (Viramma, Racine, et al. 1995).

How strange it is to realize that the blows of the iconoclast's hammer always missed their target. Are we not the inheritors of all the iconoclastic gestures of our history? Of Moses striking down the Golden Calf (Halbertal and Margalit 1992)? Of Plato breaking up the shadows of the Cave to honor this highest of all the idols, the Idea—*eidon*—itself? Of Paul sending all the pagan idols packing? Of the great wars of the Byzantine era between iconoclasts and iconodules (Mondzain 1996)? Of the Lutherans deciding what should and what should not be painted (Koerner 1995)? Of Galileo shattering the antique cosmos? Of the revolutionaries tearing down the *ancien régime?* Of Marx denouncing the illusions of commodity fetishism? Of Freud turning the fetish into a stopper that closes off the horrifying discovery of what is always missing? Of Nietzsche, the philosopher with a hammer, smashing every idol, or, more accurately, tapping them gently to hear how hollow they sound? To believe the opposite, to renounce this pedigree, this prestigious genealogy, would be to accept the grave accusation of becoming archaic, reactionary, even pagan. How could such an absurd position lead to another model for politics?

First of all, "paganism," "archaism," and "reaction" are dangerous things, but only when used as foils for modernization. There is, as anthropology has been teaching us lately, no such thing as an archaic primitive culture to which one could return. This has never been anything but an exotic fantasy of reactionary racism. The same is true of paganism, and of reactionary politics, itself an invention of modernizers. "Reactionary" is a dangerous and unstable word (Hirschman 1991), but it might be understood as simply the wish to bring care and caution *back* into the fabrication of facts and to make the salutary "Beware!" audible again in the depths of the laboratories—including those of the science students. In that sense, only the modernists want to drag us back to an earlier time and an earlier settlement, and this nonmodern precaution appears commonsensical enough, perhaps even progressive—if we accept that progress means stepping into an even more entangled future, as we saw in Chapter 6.

Second, becoming nonmodern again necessarily implies a reworking of our genealogy and of our ancestry. Idolatry may have been, all along, a misplaced target for monotheism. The fight against icons may have been the wrong battle for Byzantine emperors to wage. The Protestant Reformation probably chose the wrong target in fighting Catholic piety. Irrationalism may have been the wrong target for science; commodity fetishism the wrong target for Marxism; divinity the wrong target for psychiatry; realism the wrong target for social constructivism. Each time the error is the same, and comes from *the naive belief in the others' naive belief.* The modernists have always had a hard time understanding themselves because of their iconoclasm and because of the anxiety that icon-breaking brings. To study iconoclasm anthropologically, as part of the moderns' total way of life, as their ideal psychosocial type, necessarily modifies its effect and its impact. The knife no longer has a cutting edge, the hammer is too heavy. We must rethink the will to iconoclasm, our most venerable virtue, since its targets are no longer viable: we will not modernize the world, "we" meaning the tiny cult of "nonbelievers" at the tip of the Western peninsula.

Third, and more important, setting aside the iconoclast's hammer allows us to see that we have always been involved in *cosmopolitics* (Stengers 1996). It is only through an extraordinary shrinking of the meaning of politics that it has been restricted to the values, interests, opinions, and social forces of isolated, naked humans. The great advantage of letting facts merge back into their disheveled networks and controversies, and of letting beliefs regain their ontological weight, is that politics then becomes what it has always been, anthropologically speaking: the management, diplomacy, combination, and negotiation of human and nonhuman agencies. Who or what can withstand whom or what? Thus another political model is offered, not one that seeks to add a supplement of soul, or asks citizens to adjust their values to the facts, or drags us back to some archaic tribal gathering, but one that entertains as many practical ontologies as there are factishes.

The role of the intellectual is not, then, to grab a hammer and break beliefs with facts, or to grab a sickle and undercut facts with beliefs (as in the cartoonish attempts of social constructivists), but to be *factishes*—and maybe also a bit facetious—*themselves,* that is, to *protect the diversity* of ontological status against the threat of its transformation into facts and fetishes, beliefs and things. No one is asking Jagannath to be content with his high-caste rank and to maintain the status quo. But, at the same time, no one is asking him to debunk the sacred family stones or to set the others free. In the long history of the model of the critique, we underestimated the meaning of freedom, the freedom that comes from adding human agency twice: to the fabrication of fetishes and to that of facts. We seem to have missed something along the way. It may be time to retrace our steps; the risk of appearing reactionary may be smaller than that of being modernist at the wrong time and in the wrong way.

The subject-object dichotomy has lost its ability to define our humanity because it no longer allows us to make any sense of an important little adjective: "inhuman." What is inhumanity? Look at how strange it was in the modernist era. To protect subjects from falling into inhumanity—subjectivity, passions, illusions, civil strife, delusions, beliefs—we needed the firm anchor of objects. But then objects also began generating inhumanity, so that in order to protect objects from falling into inhumanity—coldness, soullessness, meaninglessness, materialism, despotism—we had to invoke the rights of subjects and "the milk of human kindness." Inhumanity was thus always the inaccessible joker in the *other* stack of cards. Surely this cannot pass for common sense. It is certainly possible to do better, to locate inhumanity somewhere else: in the gesture that produced the subject-object dichotomy in the first place. This is what I have tried to do by suspending the anti-fetishist's urge. The green field of humanity is not far off on the other side of the fence, but close at hand, in the movement of the factish.

In the Tel Aviv Diaspora Museum one can see a medieval illumination in which Abraham's gesture, interrupted by the hand of God, aims at the helpless Isaac standing on a pedestal; the child strikingly resembles an idol about to be broken. This bloodiest of all cities is founded on an interrupted human sacrifice. Is not one of the many causes of this bloodshed the strange contradiction there is in suspending human sacrifices while carrying out the destruction of idols with self-righteousness and glee? Should we not abstain from *this* destruction of humanity too? Whose hand should restrain us before we carry out the critical gesture? Where is the ram that could be used as a substitute for the critical mode of reasoning? If it is true that we are all descendants of Abraham's suspended knife, what sort of *people* will we become when we also abstain from destroying factishes? Jagannath was left pondering: "When they touched it, we lost our humanity—they and me, didn't we? And we died. Where is the flaw of it all, in me or in society? There was no answer. After a long walk he came home, feeling dazed."

Conclusion

What Contrivance Will Free Pandora's Hope?

What have we achieved through this admittedly strange and bumpy exploration of the reality of science studies? One point at least should, by now, be established: there exists only *one* settlement, which connects the questions of ontology, epistemology, ethics, politics, and theology (see Figure 1.1). There is thus no longer much sense in pursuing in isolation questions like "How can a mind know the world outside?" "How can the public participate in technical expertise?" "How can we elevate ethical barriers against the power of science?" "How can we protect nature from human greed?" "How can we build a livable political order?" Very quickly inquiries into these matters stumble over so many aporias, since the definitions of nature, society, morality, and the Body Politic were all produced together, in order to create the most powerful and most paradoxical of all powers: a politics that does away with politics, the inhumane laws of nature that will keep humanity from falling into inhumanity.

It should also now be clear that science studies does *not* occupy a position inside this old settlement, no matter how hard the science warriors try to contain it within the narrow confines of modernism. Science studies does not say that facts are "socially constructed"; it does not spur the masses to smash their way through the laboratories; it does not claim that humans are forever cut off from the outside world and locked in the cells of their own viewpoints; it does not wish to go back to the rich, authentic, and humane premodern past. What is most bizarre to the eyes of the social scientists is that science studies is not even critical, debunking, or provocative. By shifting attention from the theory of science *to its practice**, it has simply happened, by chance, upon the frame that held together the modernist settlement. What, at the level of theory, looked like so many different and unconnected questions to be taken seriously but independently, revealed themselves, when daily practice was scrutinized, as being tightly intertwined.

Then everything followed quite logically. Since an enormous number of conundrums had been attached to the theory of science, once we shifted our attention to practice all those classical topics became shaky as well. Hence the bouts of megalomania that, from time to time, seem to agitate science studies—some of them probably emanating from my own word processor. Is it our fault if so many cherished values—from theology to the very definition of the social actor, from ontology to the very conception of what a mind is—have been hooked upon a theory of science that a few months of empirical inquiries are enough to put into serious doubt? This does *not* mean that all these issues are not important, or that these values should not *be defended;* on the contrary, it means that they have to be fastened with a rather sturdier nail and joined to the fate of somewhat loftier ends.

I am well aware that the most contentious aspect of this search for an alternative to the old settlement is having done away with the subject-object dichotomy altogether. Since the beginning of modernity, philosophers have tried to *overcome* this dichotomy. My claim is that we should not even try. All attempts to reuse it positively, negatively, or dialectically have failed. No wonder: it is made *not* to be overcome, and only this impossibility provides objects and subjects with their cutting edges. Through inquiries, anecdotes, myths, legends, textual studies, and more than a little bit of conceptual bricolage, I have sought in this book to offer a more plausible explanation for the stubbornness of this divide: the object that sits before the subject and the subject that faces the object are *polemical* entities, not innocent metaphysical inhabitants of the world.

The object is there to protect the subject from drifting into inhumanity; the subject is there to protect the object from drifting into inhumanity. But the protective shield of factishes has disappeared, and the Body Politic has been rendered impotent. Humaneness has become irretrievable, since it is always to be found *on the other side* of that great unbridgeable gap. Once inside this huge, solemn, and beautiful architecture, no one can say a word about objects without it being used immediately to staunch some trace of subjectivity somewhere else: one cannot utter a word about the rights of subjectivity without it being seized upon to humble the power of science or to counteract the soullessness of nature. As modernity unfolded, subjectivity and objectivity became concepts of resentment and revenge. Not a trace of their liberating youthfulness can be found in them any longer. Science has been so thoroughly politicized that neither the aims of politics nor those of the sciences have remained visible. Even their common destiny has been erased. The science wars are only the latest episode in this polemical use of objectivity—and not the last, I am afraid.

I have attempted to substitute another pair—that of humans and nonhumans—for the subject-object dichotomy, which I have left untouched. Instead of overcoming the divide, I have kept the settlement where it was and headed off in a different direction, digging occasionally *beneath* the huge megaliths when it was expedient to do so: beneath, not above. I deserve no credit for doing so, since I was simply following practice, not theory. How, for instance, could I have considered, without an enormous distortion, Pasteur as a subject facing an object, the lactic acid ferment (Chapter 4)? The very subtle process of delegation that allows Pasteur to fabricate fact would have been squashed flat if I had tried to locate it in the scenography of modernism. I would have had to answer the questions bellowed by the new Fafner and Fasolt we met in Chapter 5: "Is the ferment real *or* is it fabricated?"

And it would have been even worse if I had answered "both," because the truth—the nonmodernist truth—is that facts are neither real nor fabricated, but escape altogether from this comminatory choice invented to render the Body Politic impossible. To make it through this difficult pass, they would have needed a little help from their factishes; but these facilitators have all been broken in two by the iconoclastic gesture of the critical modernists. It is not an easy thing to escape the sway of the old predicament. If readers find this book to be crudely patched together, I respectfully ask them to remember the hundreds of fragments among which I found delegation, translation*, articulation*, and the other concepts I've tried to rehabilitate, lying on the floor, smashed to pieces, deconstructed to dust! Better to have them badly restored by a clumsy curator but in full working order than to leave them there broken and useless . . .

So we have made *some* progress. There is one modernist settlement, and there is at least one alternative to it, which does not represent its fulfillment, its destruction, its negation, or its end. This is the only thing that can be asserted with some semblance of certainty. What a lively and sustainable alternative may be, I don't know. However, if we try to replace each of the fixtures of the old settlement—the boxes of Figure 1.1—we can note some specifications for the task that lies ahead.

The easiest and quickest thing to replace will be the entire artifact of epistemology. The idea of an isolated and singular mind-in-a-vat looking at an outside world from which it is thoroughly cut off, and trying nonetheless to extract certainty from the fragile web of words spun across the perilous abyss separating things from discourse, is so implausible that it cannot hold up much longer, especially since psychologists themselves have already redistributed cognition beyond recognition. There is no world outside, not because there is no world at all, but because there is no mind inside, no prisoner of language with nothing to rely on but the narrow pathways of logic. Speaking truthfully about the world may be an incredibly rare and risky task for a solitary mind steeped in language, but it is a very common practice for richly vascularized societies of bodies, instruments, scientists, and institutions. We speak truthfully because the world itself is articulated, not the other way around. That there was once a time when a war could be waged between "relativists," who claimed that language refers only to itself, and "realists," who claimed that language may occasionally correspond to a true state of affairs, will appear to our descendants as strange as the idea of a fight over sacred relics.

Second, there is clearly a space in which the sciences can unfold without being kidnapped by Science No. 1. Scientific disciplines are born free, everywhere they are in chains. I see no reason remaining for scientists, researchers, or engineers to prefer the old settlement. Epistemology was never intended to protect them, it was always a war machine—a Cold War machine, a Science War machine. The expression "socializing nonhumans to bear upon the human collective" seems to me a perfectly acceptable, albeit clearly provisional, solution, one that shelters the practice of the sciences and respects the many vascularizations they need to thrive. It is certainly better, in any case, than being submitted to this double bind: "Be absolutely disconnected"; "Be absolutely certain of what you say in words about the world out there." That this twin injunction could have passed for common sense under the pretense of fighting "relativism" will, I am convinced, seem odd a few years from now, once circulating reference has been provided to every household, like gas, water, and electricity.

Third, and more important because it concerns many more people, the conditions of felicity for politics may begin to unfold as well, now that they needn't be constantly interrupted, shortcut, quashed, and thwarted by the continual injection of inhumane laws of nature. More exactly, nature* now appears as what it always was, namely, the most comprehensive political process ever to gather into one superpower everything that must escape the vagaries of the society "down there." An objective nature facing a culture is something entirely different from an articulation of humans and nonhumans. If nonhumans are to be assembled into a collective, it will be the *same* collective, and within the *same* institutions, as the humans whose fate the sciences have brought nonhumans to share. Instead of this bipolar power source—nature and society—we will have only one, clearly identifiable source of politics for humans and nonhumans alike, and one clearly identifiable source for new entities socialized into the collective.

The word "collective" itself at last finds its meaning: it is that which *collects us all* in the cosmopolitics envisaged by Isabelle Stengers. Instead of two powers, one hidden and indisputable (nature) and the other disputable and despised (politics), *we will have two different tasks in the same collective.* The first task will be to answer the question: How many humans and nonhumans are to be taken into account? The second will be to answer the most difficult of all questions: Are you ready, and at the price of what sacrifice, to live the good life together? That this highest of political and moral questions could have been raised, for so many centuries, by so many bright minds, for *humans only* without the nonhumans that make them up, will soon appear, I have no doubt, as extravagant as when the Founding Fathers denied slaves and women the vote.

The fourth and more difficult specification has to do with mastery. We have exchanged masters many times; we have shifted from the

God of Creation to Godless Nature, from there to *Homo faber*, then to structures that make us act, fields of discourse that make us speak, anonymous fields of force in which everything is dissolved—but we have not yet tried *to have no master at all*. Atheism, if by that we mean a general doubt about mastery, is still very much in the future; and so is anarchism, in spite of the disingenuousness of its beautiful slogan "neither god nor master," since it always had one master, man!

Why always replace one commander with another? Why not recognize once and for all what we have learned over and over again in this book: that action is slightly overtaken by what it acts upon; that it drifts through translation; that an experiment is an event which offers slightly more than its inputs; that chains of mediations are not the same thing as an effortless passage from cause to effect; that transfers of *in*formation never occur except through subtle and multiple *trans*formations; that there is no such thing as the imposition of categories upon a formless matter; and that, in the realm of techniques, no one is in command—not because technology is in command, but because, truly, *no one*, and *nothing* at all, is in command, not even an anonymous field of force? To be in command, or to master, is a property of neither humans nor nonhumans, nor even of God. It was thought to be a property of objects and subjects, except that it never worked: actions always overflowed themselves, and gnarled entanglements always ensued. The ban on theology, so important in the staging of the modernist predicament, will not be lifted by a return to the God of Creation but, on the contrary, by the realization that there is no master at all. That religion too was seized by modernists as oil for their political war machine, that theology debased itself by agreeing to play a role in the modernist settlement and betrayed itself even to the point of talking about nature "out there," the soul "in there," and society "down there," will, I hope, serve as a source of bewilderment for the next generation.

It is certainly in the forward movement of time's arrow that the future settlement can do better than the modernist one. History was never at ease in the house of modernity. Either, as we saw in Chapter 5, it had to be limited to humans, nature out there escaping it altogether, or, as we saw in Chapter 6, it had to appear under the deeply improbable guise of progress, progress itself being conceived as an *increase in detachment* that freed the objectivity of nature, the efficiency of technology, and the profitability of the market from the imbroglios of a more entangled past. Detachment! Who could now believe for a second that science, technology, and the market lead us forward to less entanglement, fewer imbroglios than in the past? No, the parenthesis of progress is now coming to a close—but, contrary to the doubts that beset the postmodern sensibility, there is no need to despair, no need even to abandon the arrow of time.

There is a future, and it does differ from the past. But where once it was a matter of hundreds and thousands, now millions and billions have to be accommodated—billions of people, of course, but also billions of animals, stars, prions, cows, robots, chips, and bytes. The only feature that kept time moving forward in modernism and made it suspend itself in postmodernism was the definition of object, subject, and politics, which has now been redistributed. That there was a decade when people could believe that history had drawn to a close simply because an ethnocentric—or better yet, epistemocentric—conception of progress had drawn a closing parenthesis will appear (indeed, already appears) as the greatest and let us hope the last outburst of an exotic cult of modernity that has never been short on arrogance.

Unfortunately, as we have learned so painfully in this century, wars have devastating effects, since they force every camp to stoop to the level of its adversary. War has never been a situation in which to think subtle thoughts, but rather has always offered licence for taking shortcuts, seizing any expedient at hand, and riding roughshod over all the values of discussion and argumentation. The Science Wars have been no exception. Just when a long and thoughtful peace was needed to reassemble the broken factishes and to reinvent a politics of united humans and nonhumans, a call to arms was heard from the Right and from the Left, and "truth squads" were sent from campus to campus to fumigate the hornets' nests of science studies. I have nothing against a good fight, but I would like to be able to choose my terrain, my witnesses, and my weapons—I want, above all, to decide for myself what my war aims are. This is what I have tried to achieve in this book.

If I have not answered the science warriors' arguments term for term—or even cited their names—it is because the science warriors too often waste their time attacking someone who has *the same name* as mine, who is said to defend all the absurdities I have disputed for twenty-five years: that science is socially constructed; that all is discourse; that there is no reality out there; that everything goes; that science has no conceptual content; that the more ignorant one is the better; that everything is political anyway; that subjectivity should be mingled with objectivity; that the mightiest, manliest, and hairiest scientist always wins provided he has enough "allies" in high places; and such nonsense. I don't have to come to the rescue of that homonym! Let the dead bury the dead, or as my mentor Roger Guillemin used to say less grandly, "Science is not a self-cleaning oven so there is nothing you can do about the layers of artifacts incrusted on its walls."

Instead of this shadowboxing, I have decided to behave as if the science wars were a respectable intellectual issue, not a pathetic dispute over funding fueled by campus journalists. According to my own cartography, it is true indeed that everything to do with progress, value, and knowledge is at stake here. In the powerful words of Isabelle Stengers (1998), if we were really setting out to debunk science's claims to know the world out there, everyone would admit that "this means war," a world war, even—at least a metaphysical one. It is a battle that is worth fighting only if there are clearly two settlements in opposition. The modernist one, which, in my eyes at least, is now clearly behind us (though it was for many decades our most cherished source of light, defended by giants before it fell to the care of dwarfs), and another that is still in the offing. If anyone wants to wage *this* war, they will now know on what grounds I stand, what values I am ready to defend, and what simple weapons I expect to wield.

But I am pretty sure that when we meet on that front line, as with my friend who asked me the question that triggered this book, "Do you believe in reality?" we will all be without weapons, dressed in civilian clothes, since the task of inventing the collective is so formidable that it renders all wars puny by comparison—including, of course, the science wars. In this century, which fortunately is coming to a close, we seem to have exhausted the evils that emerged from the open box of the clumsy Pandora. Though it was her unrestrained curiosity that made the artificial maiden open the box, there is no reason to stop being curious about what was left inside. To retrieve the Hope that is lodged there, at the bottom, we need a new and rather convoluted contrivance. I have had a go at it. Maybe we will succeed with the next attempt.

GLOSSARY

BIBLIOGRAPHY

INDEX

Glossary

ACTOR, ACTANT: The great interest of science studies is that it offers, through the study of laboratory practice, many cases of the emergence of an actor. Instead of starting with entities that are already components of the world, science studies focuses on the complex and controversial nature of what it is for an actor to come into existence. The key is to define the actor by what it does—its performances*—under laboratory trials*. Later its competence* is deduced and made part of an institution*. Since in English "actor" is often limited to humans, the word "actant," borrowed from semiotics, is sometimes used to include nonhumans* in the definition.

ACTUALIZATION OF A POTENTIALITY: A term from the philosophy of history, especially the work of Gilles Deleuze and Isabelle Stengers. The best example is the pendulum whose movement is entirely predictable from its initial position; letting the pendulum fall adds no new information. If history is conceived in this way, there is no event*, and history unfolds in vain.

ANTI-PROGRAMS: See programs of action.

APODEIXIS: See epideixis.

ARTICULATION: Like translation*, this term occupies the position left empty by the dichotomy between the object and the subject or the external world and the mind. Articulation is not a property of human speech but an ontological property of the universe. The question is no longer whether or not statements refer to a state of affairs, but only whether or not propositions* are well articulated.

ASSOCIATION, SUBSTITUTION; SYNTAGM, PARADIGM: These two pairs of terms replace the obsolete distinction between objects and subjects. In linguistics a syntagm is the set of words that can be associated in a sentence

303

("the fisherman goes fishing with a basket" thus defines a syntagm), while a paradigm is all the words that can be substituted in a given position in the sentence ("the fisherman," "the grocer," "the baker" form a paradigm). The linguistic metaphor is generalized to formulate two basic questions: Association—which actor can be connected with which other actor? Substitution—which actor can replace which other actor in a given association?

BELIEF: Like knowledge, belief is not an obvious category referring to a psychological state. It is an artifact of the distinction between construction and reality. It is thus tied to the notion of fetishism* and is always an accusation leveled at others.

BLACKBOXING: An expression from the sociology of science that refers to the way scientific and technical work is made invisible by its own success. When a machine runs efficiently, when a matter of fact is settled, one need focus only on its inputs and outputs and not on its internal complexity. Thus, paradoxically, the more science and technology succeed, the more opaque and obscure they become.

CENTER OF CALCULATION: Any site where inscriptions* are combined and make possible a type of calculation. It can be a laboratory, a statistical institution, the files of a geographer, a data bank, and so forth. This expression locates in specific sites an ability to calculate that is too often placed in the mind.

CHAIN OF TRANSLATION: See translation.

CIRCULATING REFERENCE: See reference.

COLLECTIVE: Unlike society*, which is an artifact imposed by the modernist settlement*, this term refers to the associations of humans and nonhumans*. While a division between nature* and society renders invisible the political process by which the cosmos is collected in one livable whole, the word "collective" makes this process central. Its slogan could be "no reality without representation."

COMPETENCE: See name of action.

COMPLEX VS. COMPLICATED: This opposition circumvents the traditional opposition between complexity and simplicity by focusing on two types of complexity. One, complication, deals with series of simple steps (a computer working with 0 and 1 is an example); the other, complexity, deals with the simultaneous irruption of many variables (as in primate interactions, for example). Contemporary societies may be more complicated but less complex than older ones.

CONCRESCENCE: A term employed by Whitehead to designate an event* without using the Kantian idiom of the phenomenon*. Concrescence is not an act of knowledge applying human categories to indifferent stuff out there but a modification of all the components or circumstances of the event.

CONDITIONS OF FELICITY: An expression borrowed from the theory of speech acts to describe the conditions that must be fulfilled for an act of language to have meaning. Its opposite is conditions of infelicity. I extend the definition to regimes of articulation such as science, technology, and politics.

CONTEXT, CONTENT: Terms borrowed from the history of science to situate the familiar puzzle of internalist* vs. externalist* explanations in science studies.

COPERNICAN REVOLUTION: Introduced by Kant, this has become a cliché in philosophical writings. Originally it meant the shift from geo- to heliocentrism. Paradoxically, Kant uses it to mean not a decentering of the human position in the world but a recentering of the object around the human ability to know. The expression "counter-Copernican revolution" thus combines two metaphors, one from astronomy and one from political upheaval, to refer to the movement away from all sorts of anthropomorphism, including the sort invented by Kant. Politics does not have to be made through nature*, and objects may be freed as nonhumans from the obligation to shortcut due political process.

COSMOPOLITICS: An ancient word from the Stoics to express an affiliation to no city in particular but to humanity in general. The concept acquired a deeper meaning through its use by Isabelle Stengers to mean the new politics that is no longer framed inside the modernist settlement* of nature* and society*. There are now different politics and different cosmos.

DEMARCATION VS. DIFFERENTIATION: Normative philosophy of science has devoted much effort to finding criteria to demarcate science from parascience. To distinguish between this normative enterprise and the one of this book, I use the word "differentiation" instead. Differentiation does not require the generation of one normative distinction between science and nonscience, but allows for many differences, making possible a much finer normative judgment that does not rely on the weaknesses of the modernist settlement*.

DICTUM, MODUS: Terms of rhetoric to distinguish the part of the sentence that does not change (the dictum) from the part of the sentence that modifies (hence the name "modus") the truth-value of the dictum. In the sentence "I believe the earth is getting warmer," "I believe" is the modus.

DIFFERENTIATION: See demarcation.

ENVELOPE: An ad hoc term invented to replace "essence" or "substance" and provide actors* with a provisional definition. Instead of opposing entities and history, content* and context*, one can describe an actor's envelope, that is, its performances* in space and time. There are thus not three words, one for the properties of an entity, another for its history, and a third for the act of knowing it, but only one continuous network.

EPIDEIXIS, APODEIXIS: Terms of Greek rhetoric summarizing the whole debate between philosophers and sophists. Etymologically both mean the same thing, demonstration, but the first has drifted to refer to what the sophist does—a flourish of words—and the other to describe a mathematical or at least rigorous demonstration.

EVENT: A term borrowed from Whitehead to replace the notion of discovery and its very implausible philosophy of history (in which the object remains immobile while the human historicity of the discoverers receives all the attention). Defining an experiment as an event has consequences for the historicity* of all the ingredients, including nonhumans, that are the circumstances of that experiment (see concrescence).

FACTISH, FETISHISM: Fetishism is an accusation made by a denunciator; it implies that believers have simply projected onto a meaningless object their own beliefs and desires. Factishes, in contrast, are types of action that do not fall into the comminatory choice between fact and belief. The neologism is a combination of facts and fetishes and makes it obvious that the two have a common element of fabrication. Instead of opposing facts to fetishes, and instead of denouncing facts as fetishes, it is intended to take seriously the role of actors* in all types of activities and thus to do away with the notion of belief*.

FETISHISM: See factish.

HISTORICITY: A term borrowed from the philosophy of history to refer not just to the passage of time—1999 after 1998—but to the fact that something happens in time, that history not only passes but transforms, that it is made not only of dates but of events*, not only of intermediaries* but of mediations*.

IMMUTABLE MOBILE: See inscription.

INSCRIPTION: A general term that refers to all the types of transformations through which an entity becomes materialized into a sign, an archive, a document, a piece of paper, a trace. Usually but not always inscriptions are two-dimensional, superimposable, and combinable. They are always mobile, that is, they allow new translations* and articulations* while keeping some types of relations intact. Hence they are also called "immutable mobiles," a term that focuses on the movement of displacement and the contradictory requirements of the task. When immutable mobiles are cleverly aligned they produce the circulating reference*.

INSTITUTION: Science studies has devoted much attention to the institutions that make possible the articulation* of facts. In common usage, "institution" refers to a site and to laws, people, and customs that continue in time. In traditional sociology, "institutionalized" is used as a critique of the poor quality of overly routinized science. In this book the meaning is thoroughly positive, since institutions provide all the mediations* necessary for an actor* to maintain a durable and sustainable substance*.

INTERMEDIARY: See mediation.

INTERNAL REFERENT: See referent.

INTERNALIST EXPLANATIONS, EXTERNALIST EXPLANATIONS: In the history of science these terms refer to a largely obsolete dispute between those who claim to be more interested in the content* of science and those who focus on its context*. Although this distinction has been used for decades to settle the relations between philosophers and historians, it has been totally dismantled by science studies because of the multiple translations* between context and content.

INVISIBLE COLLEGE: An expression devised by sociologists of science to refer to the informal connections among scientists as opposed to the formal structure of university affiliations.

MATTER OF FACT: The general drift of science studies is to make matters of fact not, as in common parlance, what is already present in the world, but a rather late outcome of a long process of negotiation and institutionalization. This does not limit their certainty but, on the contrary, provides all that is necessary for matters of fact to become indisputable and obvious. To be indisputable is the end point, not the beginning as in the empiricist tradition.

MEDIATION VS. INTERMEDIARY: The term "mediation," in contrast with "intermediary*," means an event* or an actor* that cannot be exactly defined by its input and its output. If an intermediary is fully defined by what causes it, a mediation always exceeds its condition. The real difference is not between realists and relativists, sociologists and philosophers, but between those who recognize in the many entanglements of practice* mere intermediaries and those who recognize mediations.

MODERN, POSTMODERN, NONMODERN, PREMODERN: Loose terms that take on more precise meanings when the conceptions of science they entail are taken into account. "Modernism" is a settlement* that has created a politics in which most political activity justifies itself by referring to nature*. Thus any conception of a future in which science or reason will play a larger role in the political order is modernist. "Postmodernism" is the continuation of modernism except that confidence in the extension of reason has been abandoned. The "nonmodern," in contrast, refuses to shortcut due political process by using the notion of nature, and replaces the modern and postmodern divide between nature and society with the notion of the collective*. "Premodernism" is an exoticism due to the invention of belief*; those who are not enthusiastic about modernity are accused of having a culture and only beliefs, not knowledge, about the world.

MODUS: See dictum.

NAME OF ACTION: An expression used to describe the strange situations—such as experiments—in which an actor* emerges out of its trials*. The actor does not yet have an essence. It is defined only as a list of effects—or performances—in a laboratory. Only later does one deduce from these performances a competence, that is, a substance that explains why the actor behaves as it does. The term "name of action" allows one to remember the pragmatic origin of all matters of fact.

NATURE: Like society*, nature is not considered as the commonsense external background of human and social action but as the result of a highly problematic settlement* whose political genealogy is traced throughout the book. The words "nonhumans*" and "collective*" refer to entities that have been freed from the political burden of using the concept of nature to shortcut due political process.

NONHUMAN: This concept has meaning only in the difference between the pair "human-nonhuman" and the subject-object dichotomy. Associations of humans and nonhumans refer to a different political regime from the war forced upon us by the distinction between subject and object. A nonhuman is thus the peacetime version of the object: what the object would look like if it were not engaged in the war to shortcut due political process. The pair human-nonhuman is not a way to "overcome" the subject-object distinction but a way to bypass it entirely.

PARADIGM: See association.

PERFORMANCE: See name of action.

PHENOMENON: In Kant's modernist solution, a phenomenon is the meeting point of things in themselves—which are inaccessible and unknowable but whose presence is necessary to avoid idealism—and the active involvement of reason. None of these features is kept in the notion of proposition*.

PRACTICE: Science studies is not defined by the extension of social explanations to science, but by emphasis on the local, material, mundane sites where the sciences are practiced. Thus the word "practice" identifies types of studies that are exactly as far from the normative philosophies of science as they are from the usual efforts of sociology. What has been revealed through the study of practice is not used to debunk the claims of science, as in critical sociology, but to multiply the mediators* that collectively produce the sciences.

PRAGMATOGONY: A neologism invented by Michel Serres on the same template as "cosmogony" to mean a mythical genealogy of objects.

PREDICATION: A term of rhetoric and logic meaning what happens in the activity of definition when, to avoid a tautology, a term is necessarily defined through the use of another term. This entails for each definition a translation*, the one being obtained through the mediation* of the other.

PROGRAMS OF ACTION, ANTI-PROGRAMS: Terms from the sociology of technology which have been used to give technical artifacts their active and often polemical character. Each device anticipates what other actors, humans or nonhumans, may do (programs of action), but these anticipated actions may not occur because those other actors have different programs—antiprograms from the point of view of the first actor. Hence the artifact becomes the front line of a controversy between programs and anti-programs.

PROJECT: The great advantage of technology studies over science studies is that the former deals with projects that are obviously neither objects nor subjects nor any combination of the two. A large part of what is learned in the study of artifacts is then reused to study facts and their history.

PROPOSITION: I do not use this term in the epistemological sense of a sentence that is judged to be true or false (for this I reserve the word "statement"), but in the ontological sense of what an actor offers to other actors. The claim is that the price of gaining analytical clarity—words severed from world and then reconnected by reference and judgment—is greater and produces, in the end, more obscurity than granting entities the capacity to connect to one another through events*. The ontological meaning of the word has been elaborated by Whitehead.

REFERENCE, REFERENT: Terms from linguistics and philosophy that are used to define, not the scenography of words and the world, but the many practices that end up in articulating propositions*. "Reference" does not designate an external referent that will be meaningless (that is, literally without means to achieve its movement), but the quality of the chain of transformation, the viability of its circulation. "Internal referent" is a term from semiotics to mean all the elements that produce, among the different levels of signification of a text, the same difference as the one between a text and the outside world. It is connected to the notion of shifting*.

RELATIVE EXISTENCE: As a consequence of the positive meaning of relativism*, the insistence on the emergence of actors, the pragmatic and relational definition of action, and the importance given to envelopes*, it is possible to define existence not as an all-or-nothing concept but as a gradient. This allows for much finer differentiations* than the demarcation between existence and nonexistence. It also makes it possible to avoid using the notion of belief*.

RELATIVISM: This term does not refer to the discussion of the incommensurability of viewpoints—which should be called absolutism—but only to the mundane process by which relations are established between viewpoints through the mediation* of instruments. Thus insisting on relativism does not weaken the connections between entities, but multiplies the paths that allow one to move from standpoint to standpoint. Science studies has elaborated a new solution to replace the simpleminded distinction between local and universal.

SETTLEMENT: Shorthand for the "modernist settlement," which has sealed off into incommensurable problems questions that cannot be solved separately and have to be tackled all at once: the epistemological question of how we can know the outside world, the psychological question of how a mind can maintain a connection with an outside world, the political question of how we can keep order in society, and the moral question of how we can live a good life—to sum up, "out there," "in there," "down there," and "up there."

SHIFTING IN, SHIFTING OUT, SHIFTING DOWN: Terms from semiotics to designate the act of signification through which a text relates different frames of reference (here, now, I) to one another: different spaces, different times, different characters. When the reader is sent from one plane of reference to another, it is called shifting out; when the reader is brought back to the original plane of reference, it is called shifting in; when the matter of expression is entirely changed, it is called shifting down. These shifts result in the production of an internal referent*, a depth of vision, as if one is dealing with a differentiated world.

SOCIETY: The word does not refer to an entity that exists in itself and is ruled by its own laws by opposition to other entities, such as nature; it means the result of a settlement* that, for political reasons, artificially divides things between the natural and the social realms. To refer not to the artifact of society but to the many connections between humans and nonhumans*, I use the word "collective*" instead.

SUBSTANCE: This word designates what "lies beneath" properties; science studies has not attempted to do away with the notion of substance altogether but to create a historical and political space in which newly emerging entities are slowly provided with all their means, all their institutions*, to be slowly "substantiated" and rendered durable and sustainable.

SUBSTITUTION: See association.

SYNTAGM: See association.

SYNTHETIC A PRIORI JUDGMENT: An expression employed by Kant to solve the problem of the fecundity of knowledge while at the same time insisting on the primacy of human reason in the shaping of knowledge. As opposed to analytical *a priori* judgments, which are tautological and sterile, and synthetic *a posteriori* judgments, which are fecund and merely empirical, these judgments are simultaneously *a priori* and synthetic. When one deals with articulated propositions* this classification becomes obsolete, since neither fecundity—events*—nor logic has to be allocated between the object and subject poles.

TRANSLATION: Instead of opposing words and the world, science studies, by its insistence on practice*, has multiplied the intermediary terms that focus on the transformations so typical of the sciences; like "inscription*" or "articulation*," "translation" is a term that crisscrosses the modernist settlement*. In its linguistic and material connotations, it refers to all the displacements through other actors whose mediation is indispensable for any action to occur. In place of a rigid opposition between context* and content*, chains of translation refer to the work through which actors modify, displace, and translate their various and contradictory interests.

TRIALS: In their emerging state, actors* are defined by trials, which can be experiments of various sorts in which new performances* are elicited. It is through trials that actors are defined.

Bibliography

Alder, K. 1997. *Engineering the Revolution: Arms and Enlightenment in France, 1763–1815*. Princeton: Princeton University Press.

Apter E., and W. Pietz, eds. 1993. *Fetishism as Cultural Discourse*. Ithaca: Cornell University Press.

Aquino, P. d., and J. F. P. d. Barros. 1994. "Leurs noms d'Afrique en terre d'Amérique." *Nouvelle revue d'ethnopsychiatrie* (24): 111–125.

Beck, B. B. 1980. *Animal Tool Behavior: The Use and Manufacture of Tools by Animals*. New York, London: Garland STPM Press.

Beck, U. 1995. *Ecological Politics in an Age of Risk*. Cambridge: Polity Press.

Bensaude-Vincent, B. 1986. "Mendeleev's Periodic System of Chemical Elements." *British Journal for the History and Philosophy of Science* 19: 3–17.

Bloor, D. [1976] 1991. *Knowledge and Social Imagery*, 2d ed. with new foreword. Chicago: University of Chicago Press.

Callon, M. 1981. "Struggles and Negotiations to Decide What Is Problematic and What Is Not: The Sociologics of Translation." In K. D. Knorr, R. Krohn, and R. Whitley, *The Social Process of Scientific Investigation*, 197–220. Dordrecht: Reidel.

Cantor, M. 1991. "Félix Archimède Pouchet scientifique et vulgarisateur." Thèse de doctorat, Université d'Orsay.

Cassin, B. 1995. *L'effet sophistique*. Paris: Gallimard.

Chandler, A. D. 1977. *The Visible Hand: The Managerial Revolution in American Business*. Cambridge, Mass.: Harvard University Press.

Conant, J. B. 1957. *Pasteur's Study of Fermentation*. Cambridge, Mass.: Harvard University Press.

De Brosses, C. 1760. *Du Culte des dieux fétiches*. Paris: Fayard, Corpus des Oeuvres de Philosophie.

De Waal, F. 1982. *Chimpanzee Politics: Power and Sex among Apes*. New York: Harper and Row.

Descola, P., and G. Palsson, eds. 1996. *Nature and Society: Anthropological Perspectives.* London: Routledge.

Despret, V. 1996. *Naissance d'une théorie éthologique.* Paris: Les Empêcheurs de penser en rond.

Détienne, M., and J. P. Vernant. 1974. *Les Ruses de l'intelligence. La métis des Grecs.* Paris: Flammarion Champs.

Eco, U. 1979. *The Role of the Reader: Explorations in the Semiotics of Texts.* London: Hutchinson; Bloomington: Indiana University Press.

Eisenstein, E. 1979. *The Printing Press as an Agent of Change.* Cambridge: Cambridge University Press.

Ezechiel, N., and M. Mukherjee, eds. 1990. *Another India: An Anthology of Contemporary Indian Fiction and Poetry.* London: Penguin.

Farley, John. 1972. "The Spontaneous Generation Controversy—1700–1860: The Origin of Parasitic Worms," *Journal of the History of Biology* 5: 95–125.

——— 1974. *The Spontaneous Generation Controversy from Descartes to Oparin.* Baltimore: Johns Hopkins University Press.

Frontisi-Ducroux, F. 1975. *Dédale. Mythologie de l'artisan en Grèce Ancienne.* Paris: Maspéro–La Découverte.

Galison, P. 1997. *Image and Logic: A Material Culture of Microphysics.* Chicago: University of Chicago Press.

Geison, G. 1974. "Pasteur." *Dictionary of Scientific Biography,* ed. C. Gillispie, 351–415. New York: Scribner.

——— 1995. *The Private Science of Louis Pasteur.* Princeton: Princeton University Press.

Goody, J. 1977. *The Domestication of the Savage Mind.* Cambridge: Cambridge University Press.

Greimas, A. J., and J. Courtès, eds. 1982. *Semiotics and Language: An Analytical Dictionary.* Bloomington: Indiana University Press.

Hacking, I. 1983. *Representing and Intervening: Introductory Topics in the Philosophy of Natural Science.* Cambridge: Cambridge University Press.

——— 1992. "The Self-Vindication of the Laboratory Sciences." In *Science as Practice and Culture,* ed. A. Pickering, 29–64. Chicago: University of Chicago Press.

Halbertal, M., and A. Margalit. 1992. *Idolatry.* Cambridge, Mass.: Harvard University Press.

Heidegger, M. 1977. *The Question Concerning Technology and Other Essays.* New York: Harper and Row.

Hirschman, A. O. 1991. *The Rhetoric of Reaction: Perversity, Futility, Jeopardy.* Cambridge, Mass.: Harvard University Press.

Hirshauer, S. 1991. "The Manufacture of Bodies in Surgery." *Social Studies of Science* 21(2): 279–320.

Hughes, T. P. 1983. *Networks of Power: Electrification in Western Society, 1880–1930*. Baltimore: Johns Hopkins University Press.

Hutchins, E. 1995. *Cognition in the Wild*. Cambridge, Mass:. MIT Press.

Iacono, A. 1992. *Le fétichisme. Histoire d'un concept*. Paris: PUF.

James, W. [1907] 1975. *Pragmatism and The Meaning of Truth*. Cambridge, Mass.: Harvard University Press.

Jones, C., and P. Galison, eds. 1998. *Picturing Science, Producing Art*. London: Routledge.

Jullien, F. 1995. *The Propensity of Things: Toward a History of Efficacy in China*. Cambridge, Mass.: Zone Books.

Koerner, J. L. 1995. "The Image in Quotations: Cranach's Portraits of Luther Preaching." In *Shop Talk: Studies in Honor of Seymour Slive*, 143–146. Cambridge, Mass.: Harvard University Art Museums.

Kummer, H. 1993. *Vies de singes. Moeurs et structures sociales des babouins hamadryas*. Paris: Odile Jacob.

Latour, B., and P. Lemonnier, eds. 1994. *De la préhistoire aux missiles balistiques—l'intelligence sociale des techniques*. Paris: La Découverte.

Latour, B., P. Mauguin, et al. 1992. "A Note on Socio-technical Graphs." *Social Studies of Science* 22(1): 33–59; 91–94.

Law, J., and G. Fyfe, eds. 1988. *Picturing Power: Visual Depictions and Social Relations*. London: Routledge.

Lemonnier, P., ed. 1993. *Technological Choices: Transformation in Material Cultures since the Neolithic*. London: Routledge.

Leroi-Gourhan, A. 1993. *Gesture and Speech*. Cambridge, Mass.: MIT Press.

Lynch, M., and S. Woolgar, eds. 1990. *Representation in Scientific Practice*. Cambridge, Mass.: MIT Press.

MacKenzie, D. 1990. *Inventing Accuracy: A Historical Sociology of Nuclear Missile Guidance*. Cambridge, Mass.: MIT Press.

McGrew, W. C. 1992. *Chimpanzee Material Culture: Implications for Human Evolution*. Cambridge: Cambridge University Press.

McNeill, W. 1982. *The Pursuit of Power: Technology, Armed Force and Society since A.D. 1000*. Chicago: University of Chicago Press.

Miller, P. 1994. "The Factory as Laboratory." *Science in Context* 7(3): 469–496.

Mondzain, M.-J. 1996. *Image, icône, économie. Les sources byzantines de l'imaginaire contemporain*. Paris: Le Seuil.

Moore, A. W., ed. 1993. *Meaning and Reference*. Oxford: Oxford University Press.

Moreau, R. 1992. "Les expériences de Pasteur sur les générations spontanées. Le point de vue d'un microbiologiste. Première partie: la fin d'un mythe; Deuxième partie: les conséquences." *La vie des sciences* 9(3): 231–260; 9(4): 287–321.

Mumford, L. 1967. *The Myth of the Machine: Technics and Human Development.* New York: Harcourt, Brace and World.

Nathan, T., and I. Stengers. 1995. *Médecins et sorciers.* Paris: Les Empêcheurs de penser en rond.

Novick, P. 1988. *That Noble Dream: The "Objectivity Question" and the American Historical Profession.* Cambridge: Cambridge University Press.

Nussbaum, M. 1994. *Therapy of Desire: Theory and Practice in Hellenistic Ethics.* Princeton: Princeton University Press.

Ochs, E., S. Jacoby, et al. 1994. "Interpretive Journeys: How Physicists Talk and Travel through Graphic Space." *Configurations* 2(1): 151–171.

Pestre, D. 1984. *Physique et physiciens en France, 1918–1940.* Paris: Editions des Archives Contemporaines.

Pickering, A. 1995. *The Mangle of Practice: Time, Agency, and Science.* Chicago: University of Chicago Press.

Ruellan, A., and M. Dosso. 1993. *Regards sur le sol.* Paris: Foucher.

Schaffer, S. 1997. "Forgers and Authors in the Baroque Economy." Paper presented at the meeting "What Is An Author?" Harvard University, March.

——— 1992. "A Manufactory of OHMS, Victorian Metrology and Its Instrumentation." In *Invisible Connections,* ed. R. Bud and S. Cozzens, 25–54. Bellingham, Wash.: SPIE Optical Engineering Press.

——— 1994. "Empires of Physics." In *Empires of Physics,* ed. R. Staley. Cambridge: Whipple Museum.

Serres, M. 1987. *Statues.* Paris: François Bourin.

——— 1993. *L'origine de la géométrie.* Paris: Flammarion.

——— 1995. *The Natural Contract,* trans. E. MacArthur and W. Paulson. Ann Arbor: University of Michigan Press.

Shapin, S., and S. Schaffer. 1985. *Leviathan and the Air-Pump: Hobbes, Boyle, and the Experimental Life.* Princeton: Princeton University Press.

Star, S. L., and J. Griesemer. 1989. "Institutional Ecology, 'Translations,' and Boundary Objects: Amateurs and Professionals in Berkeley's Museum of Vertebrate Zoology, 1907–1939." *Social Studies of Science* 19: 387–420.

Stengers, I. 1993. *L'invention des sciences modernes.* Paris: La Découverte.

——— 1996. *Cosmopolitiques,* tome 1: *La Guerre des sciences.* Paris: La Découverte et Les Empêcheurs de penser en rond.

——— 1998. "The Science Wars: What about Peace?" In Baudoin Jurdant, ed., *Impostures intellectuelles. Les malentendus de l'affaire Sokal,* 268–292. Paris: La Découverte.

Strum, S. 1987. *Almost Human: A Journey into the World of Baboons.* New York: Random House.

Strum, S., and B. Latour. 1987. "The Meanings of Social: From Baboons to Humans." *Information sur les Sciences Sociales/Social Science Information* 26: 783–802.

Tufte, E. R. 1983. *The Visual Display of Quantitative Information.* Cheshire, Conn.: Graphics Press.

Viramma, J. Racine, and J.-L. Racine. 1995. *Une vie paria. Le rire des asservis, pays tamoul, Inde du Sud.* Paris: Plon-Terre humaine.

Weart, S. 1979. *Scientists in Power.* Cambridge, Mass.: Harvard University Press.

Whitehead, A. N. [1929] 1978. *Process and Reality: An Essay in Cosmology.* New York: Free Press.

Index

MALS 2001

228 – NB 2010 NJM.
283 God / Whitehead on
297 are you ready to live the good life together?